First Edition

WRITE
PRESENT·
CREATE

SCIENCE COMMUNICATION
FOR UNDERGRADUATES

By Mary Poffenroth

San Jose State University

WritePresentCreate.com

cognella®
academic publishing

Bassim Hamadeh, CEO and Publisher
Michael Simpson, Vice President of Acquisitions
Jamie Giganti, Senior Managing Editor
Jess Busch, Senior Graphic Designer
Kristina Stolte, Acquisitions Editor
Michelle Piehl, Project Editor
Alexa Lucido, Licensing Coordinator
Sean Adams, Interior Designer

First published in the United States of America in 2016 by Cognella, Inc.

Cover image copyright © Depositphotos/markovka.

Printed in the United States of America

ISBN: 978-1-63487-303-1 (pbk) / 978-1-63487-304-8 (br)

www.cognella.com   800-200-3908

CONTENTS

WRITE!
Written Communication in Science

PRESENT!

Oral Communication in Science

CREATE!

Visual Communication in Science

Acknowledgments

I'm a first generation college student. I have made tons of mistakes, and looking back, I would have loved for someone to give me a little advice along the way—even if I thought I didn't need it at the time. This book is a reflection of what I tell my students when they struggle on their projects and what I would have wanted my younger self to know. This is not meant to be an exhaustive guide to all communication, but a hyper-focused helper to give you the extra support you need to create great work. In this guide, you will find a few short stories from students across the United States who have also struggled with communication. Communication is essential to our humanity, yet we find it hard to say and write the words we need to, or create the images that connect us to each other. I urge you to keep communicating and creating thoughtful works throughout your education and for the rest of your life.

I first want to thank the entire team at Cognella Academic Publishing. My deepest gratitude goes to Kristina Stolte and Bassim Hamadeh for believing in this project and Michelle Piehl for being such a skilled and caring ringmaster, making sure all the moving parts were expertly executed. Thank you to Jess Busch for bringing the cover design to life, Dani Skeen for helping me turn a 72-word title into something that would actually fit on the cover and Katherine Huotari for your last-minute, save-the-day work on the keyword callouts.

Writing a book is like any grand adventure; the protagonist cannot do it alone. I dedicate this, my first solo book title, to my mother, Helen Poffenroth, who passed away when I first began the project. To my Godparents, Elton and Suzanne Selph, who have tirelessly loved and supported me throughout my entire life and continue to be a foundation to which I can always call home, no matter where my adventures take me. To my partner, Daniel Dobrzensky, for helping me edit the manuscript at midnight and talking me down from the "I can't do this" ledge more than a few times. Your love, partnership, and unwavering support got this book across the finish line.

I am one of those extraordinarily lucky people who has been able to create a family of best friends from across the world. What we don't share in family genetics we make up for in our dedication and friendship to each other. I am deeply grateful to Elaine Lennox and Laura Brekke, whose inexhaustible well of advice and guidance helped me bring this book to fruition from negotiation to planning to editing. To Crystal Coleman for her expertise in social media. And, finally, to Deliah Rae Taylor, Jenny Lundberg Sonawala, and Michelle McElroy Moore: I thank you for your constant and generous cheerleading, encouragement and laughter. All of you immensely important ladies have shown me strength and love throughout this process and in life.

WRITE!

Written Communication in Science

1 What is a Scientific Paper?

A Scientific Tone

Scientific tone is something that can easily elude students. In a world where we are constantly bombarded with informal literature, it can be hard not to emulate that tone in our writing. A scientific tone is more formal than a conversational tone. It is used to convey information in a professional manner. Science, at its heart, is about reputability. Having a more formal tone can help to contribute to the respect your piece will receive from the reader. That said, just because it sounds fancy doesn't mean it's true. Conversely, an informal tone doesn't mean the text isn't full of factual information. However, since the goal of this guide is to make you successful in your general education science class, it is better to err on the side of being more formal than less.

Here are some examples of the same content written in a scientific tone and an informal, conversational tone.

Example 1

Scientific tone: The brackish water of the Mediterranean ranges from a high of 31°C (88°F) to a low of 5°C (41°F) during the summer, dependent on geographic location during sampling.

Conversational tone: The sparkling blue waters of the Mediterranean Sea have a wide range of average temperatures during the summer, although you wouldn't be able to tell that from all the tourists in the water! Depending where you choose to look, the Mediterranean can be anywhere from an almost too hot 31°C (88°F) to a way too cold 5°C (41°F).

Example 2

Scientific tone: Sand, generally comprised of weathered rocks and shells, varies depending on geographic location. Differences can be attributed to the variation of hard-shelled invertebrate species present, as well as the average rate of tidal action.

Conversational tone: As any world traveler knows, beaches across the world all have different sand. What causes such a large range, from soft and white to dark and rocky? Well, it all depends on what types of rocks and hard-shelled creatures live there, as well as how rough the water is.

Science writing for higher education should not sound like a magazine article, a tweet, or a blog. Although you're not expected to sound like an expert at this stage of your academic career, it can't hurt to try.

Common Types of Scientific Papers

Science communication, like all forms of communication in the 21st century, has taken on new meaning for today's college student. Unprecedented access and abundance force students to have a discerning eye when choosing which resources to use and which to ignore. Science is inherently connected to our lives in every way. Whether or not you plan to become a full-time scientist, science is an important part of our lives. We will vote on scientifically based ethical issues like cloning; we will decide on appropriate steps to take on climate change, and which chemicals we allow in our homes and our bodies. Global citizens having, at minimum, a basic scientific understanding directly influences human health and the well-being of our planet. At some point we have all used the Internet to self-diagnose, usually resulting in misdiagnosis, panic, and unnecessary stress. Having a basic understanding of scientific principles helps us to decide whom to trust and find the quality resources we need. Good research practices help prevent us from falling into ignorance and increase our ability to live happy, healthy lives. Basically, science is just awesome!

> **Reputable Source: A trusted, honorable, respectable source**

AUDIO

Auditory delivery of science communication can be a great way to help share information. Although it is not the best to use for paper citations, it can be great way to find leads on additional resources. Science uses the same audio delivery channels as other genres. Podcasts, live/streaming webcasts, audiobooks, recorded lectures, or recordings of previous live presentations/speaker events are the most common formats used. If you are struggling for inspiration on your topic, check out audio resources from

> **Auditory:** Anything related to our sense of hearing

experts in that field. They can be an abundant resource for background information or what is new and novel in the field.

VIDEOS/ANIMATIONS

Videos and animations come in all different varieties. Some are funny, some are informative, some tell a story, and some are just the facts. Videos and animations can help illuminate concepts around your topic and improve your understanding of the subject as a whole. They can give you a glimpse into the world of working scientists and show you worlds beyond your imagination. You can get a front row seat to the world's most renowned experts and watch those experts debate each other on your topic. Videos and animations allow us a learning tool much more vibrant and engaging than many books. However, they are very hard to cite in a scientific assignment. I recommend using reputable scientific videos and animations to help you understand your topic more deeply or give you ideas on where to get even better literature resources.

WRITTEN WORKS

Periodicals/Editorials: Periodicals are written works that are generally mass produced on a regular basis: daily, weekly, monthly, or quarterly. These can be magazines, newspapers, newsletters, or reviews. In the case of editorials, sometimes they can also be opinion, or OpEd, pieces. Be extremely cautious when using opinion pieces. As always, have a discerning eye for your source. Is the writer an expert, a journalist, or a contributing community member? Everyone has a right to his or her opinion, but not everyone's opinion is supported with facts. Never forget, science is about what is actually happening, to the best of our ability to observe.

Literature Review Papers: A literature review paper is a type of written work that brings together multiple reputable sources into one cohesive report on the topic. Rarely are they exhaustive, but they can be a great way to find additional leads to other quality sources. Literature review papers can also help to give you some background information on the topic and can be used as a citation source in your final original work.

Primary Research Articles/Peer-Reviewed Journal Articles: Since there is a deeper discussion of primary/peer-reviewed journal articles in Chapter 2, I will only briefly mention them here. Primary research articles, also known as peer-reviewed journal articles, are written by the actual research scientists about their methods and results. Since this type of writing is written by those who are directly involved in the research process, it is considered the gold standard of scientific literature. These written works are published in journals that have been approved by a panel of expert peers.

Dissertations: Dissertations are usually the research culmination of a master's or doctoral degree. They can be published in journals or published in their entirety by the university directly. Many times, they hold valuable cutting-edge research. But often, they hold only minimal value to a non major. At this stage, I recommend you stick to the expert sources of information. There should be ample amounts of these sources available on your chosen topic. Of course, if you find a professionally published dissertation paper that supports your topic, feel free to use it.

Conference Proceedings: Conferences bring together researchers, industry professionals, academics, and students to share their work and learn about what is new in their field. Almost every discipline will have conferences. Conference proceedings are usually transcripts—sometimes audio/visual recordings—of the speakers who presented at a particular conference. Although they can have some great information on what is new and novel, they can be difficult to cite. If you happen to find a conference proceeding that has the perfect information for you, take an extra step and search by the speaker's name. Most likely, if they are sharing their research at a conference, they have already published it somewhere. Use this professional publication for your citation, not the original conference proceedings.

Official Reports: There are many reasons reports are published. Sometimes they are internal reports and sometimes they are meant for public viewing. Most government agencies that conduct scientific research are required to publicly report their findings at some point, since public funds are being used. Some institutions publish reports about their research to fulfill a funding requirement or as a part of their community engagement commitment. Since most reports are not formally published by a publishing house or society, citing them will be a little more difficult. However, as long as it is a reputable institution or government agency, you can usually trust that the information has withstood some level of peer review before being released to the public.

Professionally Published Books: Great books stand the test of time. When the magazines have been recycled and the Web article URLs have been deleted, books will remain. Books are an excellent resource to use, but just like any other type of communication, not all are created equal. When choosing to trust or not trust a book, look at the publisher, the author, and the date. Be wary of self-published books. When it comes to science, it is all about reputation and what makes that author qualified to write on the topic. With the ease of self-printing and publication, anyone can write a book. So, focus your attention on those books that are an easy win for reputability.

Web Articles (or eZines): Web articles are written entirely for an Internet audience and are not published anywhere else. A Web article can still be reputable, but should not be confused with a published, printed article. A printed article is something that has been published in a physical form: a magazine, journal, newspaper, etc. Web articles can be written and distributed very quickly, which makes them perfect for breaking news. You want to be very discerning when you use web articles because they run the gamut in terms of reputability. Publishing to the Internet is fast, easy, and many times very cheap or free. Check out Chapter 2 for tips on discerning which Web resources are reputable and which are not.

The Importance of Science Writing

Did you come into your science class wondering why the university is forcing you take a science? Do you feel apprehensive about science? Take a moment to think about why you have an aversion to science. Maybe you don't (woo-hoo!). Maybe you are reading this and thinking, "Hey, this does not apply to me at all!" If so, great! However, it has been my experience that many students walk into their undergraduate general education science courses with some amount of fear and dread.

Of course, I am more than a little biased. I love science! I see the value of science education everywhere I look. Knowing how the world works, knowing how our bodies work, and dreaming of innovative ways to exist in our universe is all science. To say you hate science is to say you hate breathing, sunshine, and kittens. And really, who hates kittens? Effective science communication supported by fact is essential to our ability to be the best humans we can possibly be. The goal of this guide is to help you become an even better communicator, and maybe this class will even inspire you to put a little (or a lot) more science into your everyday life.

2 Reputable Resources

Reputable sources are the foundation of any good scientific paper. Whether it's in biology, chemistry, or physics, if you want to submit a strong paper that will give you the best chance at a high score, you must support your ideas with evidence. Today, there are more sources of information available than ever before in human history, but not all of them are good ones. Anytime you submit a written work, whether for publication or for academic credit, you are putting your name and your reputation behind it. You want the best possible information available to support your ideas, hypotheses, and topic points. Submitting a paper with evidence from disreputable sources is not only unwise, but could cost you your grade and good reputation. Let's begin by categorizing the different types of published information. In this guide, we will review primary, secondary, tertiary, and Web-only literature. Please note that the word *literature* for our purposes is used to refer to *any* written work, not just the fancy books from your Studies of Victorian Literature class.

Primary, Secondary, and Tertiary Sources: What's the Difference?

PRIMARY SOURCES

Primary literature includes peer-reviewed journal articles, literature review articles, summaries, white papers, expert interviews, personal experience essays, and various other writings published by original researchers. For your project, it's best to focus on high-quality secondary sources and peer-reviewed journal articles. Peer-reviewed journal articles—hereinafter called PRJs—are written by researchers and reviewed by independent experts in their field before publication. PRJs undergo many draft revisions before being submitted to a reputable journal, and only the best research papers are approved for publication. It can take a few months to over a year for a peer-reviewed journal article to be published after the final draft has been written.

The other types of primary literature mentioned above may be helpful in providing evidence to support your paper's thesis, but be cautious. Before citing any source, you want to be completely confident in its reputability. *See Section B for more on reputable resources.*

Peer-Reviewed Journals: Peer-reviewed journal articles are the gold standard for scientific writing because scientific researchers and their teams personally write them. The information contained within this type of writing is not filtered through a journalist or textbook author; therefore, what you're reading is exactly what the researchers want to convey with their work. The upside of a PRJ is that when published in a reputable journal, you can be confident that the information presented is the best available at the time of publication. Of course, science continually changes, so new evidence, new technology, and new discoveries may occur that change what we know after publication. This is why it's important to review literature published on a wide

> **Evidence**: Specific verifiable facts that furnish proof

range of dates on your topic. The downside of primary literature is that the language is very technical, can be chock-full of complicated terminology, and will have sections beyond the scope of what you need for your project. However, primary literature is still your best option for the highest quality information. Just remember to use it as the tool it is, and not try to read it like the latest sci-fi or young adult novel.

Peer-reviewed journals are not always perfect. The Wakefield (1998) article linking autism to childhood vaccines was later retracted due to a lack of scientifically and ethically sound practices. Nothing, and no one, is perfect, but PRJs are still the best source of information in the sciences.

PRJs are usually in the same general format, with some minor differences among journal publications. The usual format is:

> **Abstract:** The 30-second elevator speech of the paper.
>
> **Introduction:** What is the problem being addressed in this paper?
>
> **Background and Significance:** Why is this problem important? What has already been done in order to solve this problem?
>
> **Materials and Methods:** Specifically, how did the researchers go about solving this problem? What tests/experiments were done? What were the variables, and how were they managed?
>
> **Results:** What happened? What was learned from the experiments outlined in the previous section? Statistical analysis results based upon the data collected are usually included here.
>
> **Discussion:** What do these findings mean? Does that data show a statistically significant pattern? Do the results collected support or refute the project's hypothesis? What further work must be done in order to solve the problem studied?
>
> **Literature Cited:** This is the section where you will find the full bibliographical information for every work cited within the entire paper. Format is dependent on the requirements of the journal that is publishing the paper.

For your nonmajor science course, you will want to focus your attention on the abstract, introduction, background and significance, and discussion sections. Unless you're creating your own research project or your instructor specifically asks you to review them, you can usually skip the research methods and results portions entirely.

Generally, the language found in the abstract, introduction, background and significance, and discussion sections will be a little less technical, have valuable additional resources, and give you a direct look into what the researchers were trying to convey. As previously stated, PRJs are a tool. By having a uniform structure that includes easy navigational cues like key terms in bold type and italicized subheadings, scientists—new and seasoned—can skim the work quickly to find specific sections that apply to them or decide if the work warrants a deeper reading. That said, don't just read the abstract and assume you know what's going on in the paper. The abstract gives a snapshot of what the research is about, but it is not a replacement for the entire paper. Additionally, you should never cite a source in your paper when you have only accessed or read the abstract, either from the full paper or from a database source like PubMed or Google Scholar. An abstract is not a substitution for a full paper.

How to Use a Peer-Reviewed Journal as a Tool: Before you jump into the peer-reviewed journal pond, have a clear idea of what it is you want to write about. Trying to find your hypothesis, goal, or the point of your paper within the depths of PRJs will most likely lead to frustration, and possible harm to your hairline or laptop computer. When beginning your paper, start with reputable secondary literature to get a sense of your topic and the direction you wish to take. Once you have a relatively clear idea of what you want to write about, then you can jump into the PRJs. When you have already formulated a clear hypothesis or idea, you then can search for evidence to support it within the sections listed above. Scientists don't always agree, and there may be conflicting positions on a particular research topic. Having a clear idea of the direction you wish to take will

> **Hypothesis**: A proposed explanation that is based on previous reputable evidence

make using PRJs a lot easier. For additional help with building your paper, see Chapter 3.

SECONDARY SOURCES

Secondary sources of literature are written works that are generally created by someone other than the scientists who performed the research. Most often, journalists write secondary sources, not scientists. Of course, there are always exceptions: some scientists write about their expertise for popular media, and some journalists have a science background. When using secondary sources, always ask: What qualifies the author to write about this topic? Quality secondary sources will usually weave primary literature into their story, as well as pull from sources such as personal interviews, current events, and public opinion polling. Reputable secondary literature sources can be helpful, as long as you are confident that the publisher has a high standard for fact checking and printing only the best information available at that time.

> **Thesis: A central idea, statement, or theory**

Reading reputable secondary sources at the start of your research process can help illuminate the topic in your mind and get your creative juices flowing. Although secondary literature can be much easier to read, be very cautious. Using secondary literature sources, especially those found on the Internet, can supply you with inaccurate—sometimes even blatantly false—information. *Not sure how to tell a reputable source? See Section E below.* Reputable secondary sources such as those articles found in publications like *Scientific American, Discover,* and *Popular Science* will be written in a format that is more palatable to the nonscientist. These articles, usually based on original research, give you the "who, what, why, when, where, and how" of a particular topic of interest, but deliver it in an editorial fashion. The editorial style of writing is commonly what students are most familiar with. This is the style of writing you find in magazines, newspapers, and professional blogs.

A sign that the secondary literature source you have in front of you is reputable is the inclusion of detailed references to primary literature. These citations can be a wonderful resource when it's time to hit the PRJs. After you have read through multiple reputable secondary literature pieces and you have been able to formulate a clear direction for your paper, you can then gather the PRJs referenced in the most applicable articles. This will save you from having to sift through hundreds, maybe thousands of results in a database search.

Just as in life, timing is everything in publishing. The amount of time between when the author has finished the final draft to the point of publication for secondary sources can be as little as a few minutes to a few weeks. This is why secondary sources may have excellent information for new discoveries, but also come with possible pitfalls. Being confident in a publication's reputability is of the utmost importance for secondary sources of literature.

TERTIARY SOURCES

Tertiary literature sources are generally works such as textbooks, encyclopedias, and field guides. These works are meant to stand the test of time; therefore, they are often far removed from current research. It takes months, sometimes years, from the point where the author has finished the final draft to the point of publication. Tertiary sources are meant to have a very long shelf life. Multiple pairs of eyes will be proofreading every word to ensure grammatical accuracy and fact checking all the information contained within the work. Tertiary sources are usually books, but with digital publishing becoming even more popular, they may be delivered in an eBook, free- or fee-based website, or Kindle/Nook edition.

Tertiary literature can be helpful for learning the background information on your topic. You will not find extremely current or novel information in a tertiary source. Always keep in mind the goal of your paper and/or project. Is your assignment to discuss what is currently happening in a particular field's research? Is it to deliver a background summary of a chosen topic within your class's discipline? Is it to come up with an original research project of your own? Make sure you are confident in what your instructor is asking you to do on this project before you start researching. If, after reading the materials provided to you by your instructor, you are even the tiniest bit unsure about what you're being asked to do—go talk to your instructor in person, now! The earlier you clear up your confusion, the less time you'll waste and the more focused you'll be.

WEB ONLY

For many students, it can be extremely confusing to discern between a Web-only publication and an electronic copy (eCopy) of a physical publication. A physical publication is a printed work found in hard copy form, somewhere in the world, that has been produced by a publishing house or professional printer (i.e., self-publishing). No, printing a website out on your personal printer does not count as a physical publication. A physical publication will have publishing information such as a copyright, an ISBN for books, and a volume/issue number for journals.

When you are viewing an eCopy of a printed work in a database or website, it should be treated exactly the same as if you were viewing the hard copy. These two types of written works may sound very similar, but the differences in their reputability can be astounding. When you come across a written work online, start by asking yourself a few questions to determine if it is a Web-only publication or an eCopy of a printed publication.

 ## Is it Reputable?

We all learned early on to be discerning about the source of our food, with the clichéd lesson to never accept candy from a stranger. Even though our imaginations filled with frightening images of evil clowns doling out poisonous candy, it was a wise lesson to be learned: beware of unknown sources.

> **ISBN**: An International Standard Book Number is a 10-digit or 13-digit number assigned by the publishing industry to help track, control, and facilitate publishing activities

Although poor choices in literature resources may not result in illness or death, they could result in a poor grade or the death of your good reputation. A reputable resource is trustworthy, reliable, and filled with facts based on the best available data at the time of publication. High-performing university student that you are, it's your duty to help perpetuate high-quality, factual information. The rise of effortless publishing by anyone with access to a computer or smart phone has drastically diluted the quality of information available online. Since there is no way to control the quality of what is published, it is up to you, brave defender of science, to ensure that you will only expose yourself to high-quality sources.

Start by asking these questions about the article you're reading:

who

- Who are the authors? Never use a source that does not have an author unless it's being published by an extremely reputable publication.

- What makes the author qualified to write on this particular subject?

- Does the author have an affiliation to a not-for-profit institution, such as a university or government agency?

- Does the author work for private industry? If so, it doesn't automatically mean that it isn't reputable. Just make sure to do some additional research into the motivations of the paper.

what

- What type of research is being done?

- Is the type of research being done similar to what others are doing in that particular field of study?

- Are the methods credible? For example, does the paper discuss a method of human cloning that no other researchers are able to verify?

why

- Always ask yourself the motivations of the author. Is something being sold or marketed with this particular article?

- What does the author have to gain from its publication?

- Who would profit, if anyone, from this article being published? Again, monetary profit does not automatically mean bad science. You just want to ask the question and find a good answer.

- Is the research being performed considered ethical?

when

- How recently was this published?

- Does the publication date matter for your subject? If you are studying stem cells, you bet it does! If you are studying gravitational force, maybe not.

where

- Where was this research carried out?

- Did a university, private research company, or government agency sponsor its funding?

how

- How is the article written? Is the tone formal or informal?

- Does the author seem unbiased or argumentative?

- Is the article grammatically correct?

- What evidence does the author supply to support his or her work?

- Are there fully cited external sources of information included to support the topic points?

This process may seem tedious at first, but once you get into the habit of having a keen eye for quality resources, it will become effortless.

C   Interviewing Experts

Interviewing experts in the field will serve to both increase the richness of your paper's content and your knowledge of the topic. Before deciding to cite an expert, be sure to clear it with your professor. Although many professors are supportive of using unpublished expert communications in assignments, some are not. Interviewing an expert can also help you get clarification on complex concepts you may be struggling with, regardless of your final draft choices.

FINDING EXPERTS

In order to interview an expert, you must first find an expert. Here are a few places to look.

Professors: Professors are experts in their field? Really? Of course! It is their job to pass their knowledge to the next generation. Professors are also the easiest for you to access as a student. Look on your college website for professors who specialize in similar subjects as your project, or venture beyond your campus by contacting professors in other areas.

Graduate Students: Graduate students, those pursuing a Master's degree or PhD, are usually pretty knowledgeable about their field and easier to access than most other types of experts. Most university research labs will have biography information on the graduate students who are studying under a particular professor. See who they are, what they study, and reach out to them. Most will be flattered that you considered them as a resource for your project!

Industry: As a student, most people are willing to help you because they remember being a student themselves. Reaching out to an industry professional who is working in the field of your topic may land you an interview and maybe even a great professional connection for future opportunities. Think about someone you may already know who works in the field you are interested in, and ask for a personal introduction. Even if you don't know anyone personally, most companies provide e-mail contacts on their website. Although you may not get a response, a few minutes to craft a well-written e-mail could mean a great professional connection.

Famous Experts: If your great-aunt is Dr. Jane Goodall or your neighbor is Dr. Neil deGrasse Tyson, then by all means, shoot for the stars. If you don't have a personal connection to a top expert in the field, it most likely isn't worth your time to pursue an interview for an undergraduate project. But hey, if you're really passionate about this particular expert, why not send that person an e-mail? Keep moving forward with your project; if they respond, use the unique opportunity you've just secured as delicious gravy to top the already delicious hearty dish you have created.

PREPARING FOR THE INTERVIEW

Well done! You landed an interview with an expert. You want to honor and respect your expert, since they're taking time out of their busy schedule to talk with you. Even if it is just via e-mail or phone, do your homework. Being prepared shows your interview subject that you know what you're talking about, that you value their time, and are a student dedicated to success.

Before you reach out to your expert:

- Do some basic research into what this person studies. You don't want to approach an astrophysicist to discuss your paper on the effect of groundwater pollution on red-legged frogs.

- Do not send an e-mail or make a call to an expert without finding out the person's name. Beginning an e-mail with "Dear Sir or Madam" is a quick way to get your message deleted. Formally address them by their title (Dr., Professor, etc.), if applicable, and always use the last name. In an age where many students find it acceptable to communicate informally, stand out by being professional and formal in your early communications.

- In your initial request for an interview, compliment your expert by citing specific details from their work. Keep it simple, and make sure you actually understand what you're talking about. Showing your expert that you took the time to find and read something they've written will help you get closer to an interview. Go to their website, search their LinkedIn profile, or use your newly honed research skills to find PRJs they've worked on.

- Set up a specific time and place to meet your expert. In your initial request, let them know exactly how much time you will need. Telling them how many minutes you plan to use helps them make a more informed decision on whether they can fit you into their day.

- The day before your interview, send a quick follow-up e-mail to confirm time, date, and location.

DURING THE INTERVIEW

- Be ready to walk into your interview location at least five minutes before your meeting time. This means you have already parked, checked your hair, and gone to the bathroom (not that you might be nervous before an interview or anything). Being late clearly conveys to the other person that you do not value their time, which is the worst way to start an interview.

- If you want to be perceived as a high-performing student, you need to look like one. Does this mean a three-piece suit? No. It does mean dressing as professional as your wardrobe allows. If you have no idea what professional business attire looks like, there is a plethora of resources on the Web. Hot-pink velour sweatpants and basketball shorts are not professional attire. Yes, you can still look like the creative individual you are while looking professional. I will leave the details up to you.

- If you are conducting your interview in the United States, it is the cultural norm to shake someone's hand when you first meet. There are many studies correlating your handshake to the way your new acquaintance perceives you. Having a firm—but not a Kung-Fu grip—handshake is usually the best way to go. Not sure what your handshake feels like? Ask a friend to practice with you. A great handshake is a valuable tool to have when working in the United States.

Of course, other cultures around the globe have different practices around introductory meetings. If you are interviewing an expert outside the United States, or if your expert is visiting from another country, take that extra step to research the acceptable forms of introduction.

- Recording audio or video during an interview can be a valuable way of not missing any information being said. However, you need to explicitly ask and receive permission from your expert before arriving to your interview. Don't just surprise them with a recording device and expect them to comply. Many people find recordings to be intrusive and distracting. On the other hand, many people don't mind at all. Don't assume your expert will be willing to allow recording. Asking permission when setting up the interview will help to avoid any uncomfortable situations during your meeting.

- Have paper in front of you, even if you are recording. Create a set of questions you want to ask your expert about things that could not be found on Wikipedia. Asking creative, specific, and relevant questions exemplifies your dedication to excellence.

- Stay engaged. Do not, *under any circumstances*, check your phone during an interview. Keep eye contact the entire time; even if you're bored, act like this is the most fascinating conversation you have ever had. Remember, you asked this expert for help, not the other way around. They have a million other things to do with their busy day and are here to help you. Show them the respect they deserve by giving your full, completely undivided attention during the interview.

AFTER THE INTERVIEW

- After the interview has concluded, thank your expert again for their time and ask if they would like a copy of your project upon completion.

- Later that day, or the day after at the latest, send a handwritten, hard-copy thank-you note. Not an e-mail, not a phone call, but a physical thank-you note mailed to their place of business. You might be thinking, "No one sends thank you notes through the mail anymore." Exactly. Stand out from everyone else by showing that you cared enough to make the effort. Trust me, everyone loves to get handwritten thank-you cards. You never know when you will need that professional contact again or when they may have an opportunity for you. Making the best impression possible will help ensure you leave your expert with that warm, fuzzy feeling of knowing their time was appreciated.

CITING AN INTERVIEW WITHIN YOUR PAPER

Additional details about citing an expert's personal communication in your project can be found in Chapter 5, where the Council of Science Editors (CSE) style is reviewed. Generally, this type of citation is called a Personal Communication, or Pers. Comm. You never want to pass off information as yours that came from an outside source. Always give credit where credit is due, especially to the kind expert who devoted part of their day to helping you.

Jesse Dominguez

The executive chef of Mon Ami Gabi sat, arms folded, across from me in a small French bistro table. I was nervous, of course I was nervous. With only one year of college left, I had come out to Las Vegas to stay with a friend and, hopefully, get a job for the summer. This was my first, and hopefully only, interview. After the chef and I had exchanged pleasantries and basics, he moved on to the main course, "Did you research us before coming here today?" I froze, he continued, "What is French Cuisine to you? Do you know how long we've been open? What is a warm olive jar?" I panicked. I didn't even think to do any research beforehand, so I just started spewing a gaggle of garbled words in an effort to salvage some dignity. Of course that just made things worse. After 20 minutes of struggling to effectively communicate the knowledge I did have, we respectfully shook hands and departed with a cordial farewell that said, "I don't need a fabricator of truth, I need a communicator." As I walked toward the doors of the bistro, I realized that I never really learned how to be a strong communicator in college. I was the student who would hide in the back of the class to avoid having to answer questions. I did only what was absolutely necessary during presentations, and now, that lack of communication skill was coming back on me in the real world. How would anyone know what I had learned in college if I couldn't effectively communicate it to them? I walked through the front doors of the Paris casino into the bright June sun with one final thought, "What just happened? I only applied to be a dishwasher."

Jesse Dominguez
Midwestern State University
Wichita Falls, TX

3 Starting Your Paper

A Gathering and Organizing Your Research

Research is the hallmark of a good-quality, science-based communication piece. Although there are some works based on opinion, your introductory academic project will most likely require you to find information from reputable sources. So what is the best way to begin?

Launching any project is daunting at first, but you must start in order to finish. Many people have unique ways of creating their work. Here are a few methods that may help increase your efficiency and get you into the creative mind-set.

Begin by going from general to specific. What that means for you will depend on how much you already know about your topic. Since this guide was written with the nonscience major student in mind, you most likely are not an expert on scientific topics (yet).

Jumping directly into journals will probably not be your best bet. Start by writing down a few key words about what you think you may want to write about. Give yourself lots of freedom to change your topic at this early stage. Read the assignment given to you by your instructor, and brainstorm some ideas to help meet the goals of the assignment. Brainstorming, or writing down whatever random crazy idea comes into your head, is different for everyone, but here are some ways to get more creative:

- *Get a Noodle Board:* I like the phrase "noodle it out," which is a Silicon Valley techie word for brainstorming. A noodle board is a whiteboard that you put your ideas onto. Any size will work, but I find the bigger, the better. Why is a noodle board better than paper, you ask? Well, I feel it is a subconscious message to be more freethinking with your thoughts. Unlike paper (which feels more permanent), a noodle board is dry erase. Anything that is written or drawn on it is meant to be temporary and easily erased. This helps to free us up from the confines we invent around our ability to be creative. A noodle board can help unblock a little of the anxiety you may be feeling about starting your project and get some great ideas to the surface.

- *Noodle with Friends*: Done correctly, working in a productive, supportive group of peers can help to elevate your creativity and the quality of the final product. Take a look at some of the guidelines around group work in Chapter 8 to help you and your colleagues stay focused during your brainstorming session.

- *Timed Free Writing*: You can do anything for 15 minutes. Set a timer on your phone or microwave for 15 minutes. In bold, write the assignment prompt at the top of a word-processing page. Then, for 15 minutes, just type everything that comes to mind. Do not take time to correct spelling or grammar mistakes. Don't edit what you are thinking. Just pour out everything that comes to mind, no matter how abstract, onto the page while focusing on the topic at hand. When the timer bell rings, review what you wrote. Again, at this point it is about idea creation, not a grammatically correct work of art.

- *Concept Mapping*: In the center of a white page or noodle board, write your topic or the assignment general topic within a circle. Then, using lines and other circles, start to draw connections between other words and your topic. Don't edit yourself now, just see how you can create connections between your assignment prompt, words, and ideas.

- *Brainstorm Using Images*: Do you learn better when there are engaging images involved? We all learn and think a little differently from one another. Finding what works for your unique brain will help make your study time more efficient. Try some of these methods for brainstorming in images:

- *Drawing*: Using whatever medium you choose (pencil, pen, watercolor, charcoal, or crayon), start to sketch what comes to your mind when you think about your topic. Try it even if you're the self-proclaimed inartistic type. Connecting to your topic through a creative outlet like drawing can help rattle loose some ideas for your project.

- *Image Searching*: In your Web browser bar, select "Images." Then type in your topic. What comes up? Scroll through the images that are tagged with key words similar to your topic. Although images alone are not reputable sources for citation, you're still in the exploratory phase. In this phase, it's okay to poke around all that is available to you as you get to know your topic more deeply.

Once you have some good key words, start to learn a little more about your topic. At first, it is best to start with reputable secondary literature. As you know from reading Chapter 2, the best resources for your project are primary literature sources, but at the beginning, as you're forming your specific topic, it is easier for you to start by exposing yourself to quality sources that were specifically created for nonscientists. Begin by performing a simple Internet search using your favorite Web search tool (I'm not going to say the G word here). I will, however, plug a unique search engine called Ecosia that plants trees every time you use their free search engine. Check them out at www.ecosia.com

Once you perform your Internet search, go beyond the first page of hits. Please take a look at finding reputable Internet resources in Chapter 4.

B   Building an Outline

Every good writer outlines and completes multiple drafts. This is the reality of producing quality work. Outlining, even if it is not required, helps to build scaffolding for your work. It helps to flesh out your ideas so that you can build a coherent paper.

After you have brainstormed ideas and completed some preliminary research, begin to clump things together that are similar. Then choose three major themes that you would like to include in your paper. Why three? Three is a good start because it is more than one or two, obviously, and it does not seem overwhelming. Depending on the length, you could always add more themes later on.

Think about it this way. What three ideas would you want your instructor to be able to recall after reading your paper that would make him or her feel confident that you addressed your topic? Figure out what those are and use them to guide your research.

Once you have these three main topics, start to create subtopics. Look back to your brainstorming session notes. Can you see patterns begin to emerge? What key words fit nicely under your main topics? Are there key words that don't really fit into your three main topics? If so, save them for later, but keep your outline simple to start with by having key words that naturally fit with your three main topics.

Unless you are required to submit an outline, your outline can look anyway you wish. Get creative with it, as much as you enjoy. Just make sure to keep it organized and logical.

C   The Drafting Process

Getting a passing grade on a paper written the night before is about as likely as winning the lottery or being struck by lightning. Sure, it happens once in a while, but it's highly unlikely. Professional writers and highly successful students all do multiple drafts. It is part of the process, and I can't recommend it enough.

Even if your first draft is terrible, still draft. My recommendation, since most students will be typing their drafts on a computer, is to save each draft as it is. I label each draft I write by date. For example, *CognellaDraftJuly23* for one, then *CognellaDraftAug17* for another. This way, I can always go back to a previous draft if I wish. It helps me be less stressed about making changes to previous editions. I would also highly recommend that you keep multiple copies of your drafts somewhere separate from where you are working on them. Each day, when you finish a draft, e-mail it to yourself or save it to a flash drive. I cannot tell you how many panic-stricken students I have seen come into my office that have had the only copy of their paper on a laptop that was stolen or damaged. Make multiple copies and have them in multiple places, just in case.

Kim Zaninovich

"Only 8 of you will make it to the final exam." All 12 of us looked around the room at each other on that fateful first day in Professor Maître's class. Who would be one of the 8? It was senior year, and the class was a graduate-level Journalism course in international relations. Professor Maître had covered every conflict since Korea for various prestigious publications around the world. He knew his stuff, and I knew I was in over my head, but I was determined to be one of those 8 students.

My chance came when Professor Maître asked students to volunteer to write an article based on an additional, non-mandatory reading assignment. I jumped at the chance and chose a novel called Pettibone's Law by a former Marine fighter pilot, John Keene. Professor Maître's words "make it publishable" echoed in my mind as I struggled to find a hook for my article.

It was 1994, and Googling for inspiration was not yet an option. I wanted, I needed, to capture the attention of my classmates and my Professor with this piece, but how?

I decided to find a Connecticut phone book and look for all the John Keenes. After a few unsuccessful tries, I eventually found the author! He was lovely, surprised by my call, and very willing to talk to a stranger about his work. My professor liked my initiative, and I was able to create something that felt "publishable" for my classmates. With that brief assignment, I gained two important insights that I call on to this day. First, there's always an angle; there's always a way to communicate more effectively, weave a better story, find depth where you fear there is none. Second, respect your audience. If you find yourself out of ideas, take a breath, and go back to the source to view your audience as a muse. Works every time.

Yes, it was a graduate level class. Yes, it was challenging for an undergraduate senior, but I was one of those 8 students at the final exam and passed the class with flying colors.

Kim Zaninovich
Boston University
Boston, MA

4 Online Resources

A Discerning Reputability of Websites

When you begin searching for quality resources, the Internet is usually your first stop. Consequently, it is extremely important for you to learn how to discern which websites you should trust, which you should be cautious of, and which ones you should avoid completely.

Popping key words into any Internet search engine will usually get you more results than could possibly be viewed in a lifetime. Obviously, you will not be looking into every link, but definitely go past the first few results. Most search engines will give you at least the web address and a few words that the site contains in your initial search results. Begin by looking for results that have an *.edu* or *.gov* domain extension. Although these two extensions don't guarantee quality, they are a wonderful place to start.

.EDU

A website that has the domain extension .*edu* is going to be an academic institution. However, just because a school publishes something on a website does not mean it is automatically reputable. Ask yourself these questions when deciding whether or not to use information from a .*edu* website:

1. Have you heard of this school?
2. Is it a prestigious university or an unknown high school?
3. Is what you are looking for a specialty at that institution? For example, getting information about your paper on diabetes from Johns Hopkins University is definitely a safe bet.
4. Do they offer additional links to reputable external sources?

> **Discern**: To observe, identify, and distinguish between

AN UNFAMILIAR WEBSITE

What is the quality of the Web design? Does it look like someone is constantly maintaining the site (which costs time and money), or is the overall look basic, outdated, or clearly neglected?

Is there a "last updated" date somewhere at the bottom of the home page?

Who created the website? An individual, a school, a government agency, a nonprofit group, or a private company?

Who is in control of the content? Can the content be changed by anyone, which is the case with websites like Wikipedia?

What qualifies the website creator to be publishing information on your topic? This can usually be found in the "About Us" section.

What interest does the website creator have in presenting you with this information? Are they trying to sell you something, nonpartisan, or a group with a particularly strong position?

Does the information presented seem unbiased? Science, as much as possible, attempts to learn and deliver information in an unbiased way. If a website seems argumentative and advocates strongly for a particular position, especially without providing external evidence, it appears biased.

What is the tone of the language? Does it sound informative and professional? Or sensationalistic? Does it remind you of someone trying to sell you soap on a midnight infomercial?

Here are *.edu* websites I recommend for biology:

mvz.berkeley.edu—Museum of Vertebrate Zoology at Berkeley

animaldiversity.ummz.umich.edu—Animal Diversity Web, University of Michigan

.GOV

A website with the URL extension *.gov* is a sponsored top-level domain that can only be used by United States federal and state government agencies, counties, or cities. A federal agency *.gov* website is considered a reputable website and will usually have some excellent information on your topic's background. Federal agencies will have a portion of their budget set aside for public outreach and education. This benefits you because they're writing complex scientific information in a style that is palatable for the general public.

Here are *.gov* websites I recommend for biology:

NOAA.gov—National Oceanic and Atmospheric Administration

EPA.gov—Environmental Protection Agency

Science.gov—U.S. Federal Science Open Access Gateway

Energy.gov—U.S. Department of Energy

USDA.gov—U.S. Department of Agriculture

FWS.gov—U.S. Fish and Wildlife Service

<u>Nature.NPS.gov</u>—National Park Service

.COM

After you have exhausted your *.edu* and *.gov* hits, start glancing over *.com* and *.org* websites. This is where you want to use even more scrutiny. A Web address with the domain extension *.com* could be one of a plethora of varied types of websites. Originally, *.com* meant "commerce," but today it can be associated with any type of website. With *.coms*, you want to have a very discerning eye, unless you are already familiar with the reputability of the source. For example, here are some science *.com* websites that I consider reputable and that provide quality secondary and Web-only sources:

<u>ScientificAmerican.com</u>—*Scientific American* magazine

<u>ScienceDaily.com</u>—Science Daily Research News

<u>PopularScience.com</u>—*Popular Science* magazine

<u>Science.NationalGeographic.com</u>—National Geographic Science Portal

In addition to the basic questions you always want to ask yourself when visiting a new website (listed above), you also want to pay particular attention to the motivation behind a *.com* website.

- Are the website creators trying to sell you something? This doesn't mean not to trust the source; just be aware.

  o For example, *Scientific American* is a magazine. They are essentially selling information. In order to preserve their good reputation, they must continue to provide the public with quality, reputable resources.

- Could their motivation for you to buy their product be clouding the reputability of their information?

o A scientific-looking graph showing how a new weight loss pill will make you drop ten pounds in just one week probably is not the type of data you want to include in your paper.

- Does the information presented have links to alternate, outside sources of information?

.ORG

The *.org* domain extension is generally used by an organization of some kind. This could be a for-profit or nonprofit organization. As with the above types of web sources, just because a website is funded by an organization, does not automatically mean that you should trust or distrust what they are publishing. Here, you want to pay special attention to what the goals of the organization are.

- Are the goals of the organization to further research and community engagement?

- Is it a philanthropic organization?

- How long has the organization been around?

- What do they have at stake? In other words, what do they risk by publishing incorrect information?

Here are *.org* websites I recommend for biology:

WWF.org—World Wildlife Fund

Nature.org—Nature Conservancy

WHO.org—World Health Organization

PBS.org—Public Broadcasting System

There are various other extensions out there like *.it, .uk, .idk,* and *.lol.* Always ask yourself the basic questions above when deciding on any new source of information. Don't forget: anyone can get a website and publish on the Web whatever they wish, whether it is supported by evidence or not.

 ## Databases vs. Websites

A portion of your university fees directly supports the campus library, so make sure you maximize all of your free research resources. Using search engines such as Google Scholar (www.googlescholar.com) will usually return hits with an abstract only or ask you to pay to view entire articles. Since you never want to cite a reference off its abstract alone, you need to view the full article. You can get access to millions of full-text articles through your library! Never pay for access to articles as a student. Even if there is an article that you are dying to get and your library doesn't have immediate access, most libraries can get it for you through an interlibrary loan.

Let's begin by defining what a library database is and is not. A library database:

- electronically stores digital copies of physically printed published articles.

- will provide you with search functions.

- may or may not provide full-text articles.

- may or may not contain only peer-reviewed journals.

- is not a website or considered a Web source. Even though it lives on the Internet and looks like a website, it is a storage repository for physically published articles. Think of it as a giant electronic filing cabinet. This will be vitally important when it is time to cite your sources.

SEARCHING IN A DATABASE

A more specific search will result in better results. That said, being overly specific may not get you enough information; it is a happy medium. For example, searching for "monkeys" will return 476,537 articles. Searching for "capuchin monkeys" will return 123,543 articles. Searching for "capuchin monkeys Costa Rica" will return 43,076, and searching for "capuchin monkeys Costa Rica diseases" will return ... well, you get the idea.

Also, look for "Advanced Search" features somewhere in the search bar area. This will allow you to further refine your search to specific dates, volumes, publication type, etc. If there are no advanced search features, using a Boolean phrase such as "AND," "OR," and "NOT" typed in all caps followed by a term can also help to refine your results.

There are many good databases out there, and depending on your university's subscriptions, you may or may not have access to the ones mentioned below. However, you want to look for a few key features when deciding whether a database will adequately fulfill your needs:

- Does it provide access to full-text articles?

- Which journals does the database include? This can usually be found somewhere in the "About Us" section.

- Is it free to use and download full-text articles?

Here are journal databases I recommend for biology:

- Wiley ScienceDirect

- JSTOR

- BioOne

 Using and Citing Web Sources and Multimedia

Always use caution when citing a source in your final project from a Web-only written source. In general, citing a video, animation, or audio file like a podcast is usually frowned upon because it is difficult to verify the content being referenced, and hyperlinks are quickly changed or removed. The Internet is an extremely fluid source. Unlike physical publications that will be found somewhere in the world for decades, sometimes even millennia, material published in a Web-only format has an extremely short shelf life of anywhere from a few years to as little as a few minutes. This is why you want to cite Web-only sources sparingly in your final product.

Although it has been mentioned before, it's good to mention it again. When you are looking at an electronic copy (eCopy) of a physical publication on the Internet, it is still considered a physical publication and should be cited as such. This tends to be an area where students get confused. When you find, let's say, a journal paper in JSTOR or ScienceDirect, you are looking at a digital version of a printed paper. Somewhere in the world, that paper exists in hard copy form, professionally published in a journal, and available for people to examine. You just happen to be looking at the electronic copy at the moment. Even if the website or database is taken down, the resource continues to exist in libraries and private collections.

When reviewing an eCopy, look for these characteristics of a physical publication:

- Publication date

- Authors' full names and possibly professional titles

- Name of publishing house or society

- Publication location

- Volume, issue, and/or series number

- Page numbers

5 Citation Formats Used in Science

A Citations in the Sciences

Although you may be more familiar with citing your sources using MLA or APA style, scientists generally do not utilize those methods of formatting. When writing a paper for publication, the publisher or society decides the format the author is required to use. The three most widely used style formats in science are:

- Council of Science Editors (CSE)
  councilscienceeditors.org

- American Chemical Society (ACS)
  pubs.acs.org/series/styleguide

- American Institute of Physics (AIP)
  scitation.aip.org

Since the Council of Science Editors, or CSE, style is the most widely used in biology, and I am a biologist, I will briefly cover one of their accepted styles here for an example. However, there is surplus of excellent websites available, in addition to the sites above, that can help you utilize the correct citation style for your science course. For additional resources on citations, please go to WritePresentCreate.com.

CSE has three systems of documentation, otherwise known as sub-styles. They are:

- Name–Year*

- Citation–Name

- Citation–Sequence

Let's use the CSE Name–Year Style as our working example here.

B Council of Science Editors (CSE) Style—Name-Year

There are some general guidelines you should follow when using the CSE Name-Year style of citation.

- You must cite all sources of information that did not come from your brain or common knowledge.

- If there are more than two authors in a paper, cite the source within the body of your paper as (Smith et al. 2008). Et al. means "additional authors." However, in the literature-cited section at the end of the paper, you must spell out all authors who contributed to the paper.

- If you use multiple pages from a single website, it is still considered just one website and one source.

- If you have used a reference multiple times within the same paragraph, you may cite your reference at the end of the last sentence in that paragraph. Otherwise, cite your source at the end of the sentence in which you are using someone else's ideas, data, or information.

- All citations that are referenced in the body of the paper must appear fully cited in the literature-cited section, and vice versa. In other words, your in-text citations and literature-cited section citations need to match.

- Do not include URL information for peer-reviewed articles you retrieved from an online database. As you remember from previous chapters, these are simple electronic copies of physically published journal papers.

- Each individual citation in your literature-cited section should be single spaced, placed in alphabetical order by the first author's last name, and with a double space separating individual citations.

- Never reorder a paper's sequence of authors. Being first author on a paper is a very prestigious position. Never alphabetize the authors of an individual paper.

- Left justify your literature-cited section with no hanging indents.

In-Text Citations: These are the citations you will use within the body of your paper. You may place the citation in the middle or at the end of a sentence.

If your resource has only one author:
Hummingbirds are active during the day (Jones 1984), while barn owls are nocturnal and only active at night (Smith 1982).

If your resource has two authors:
Hummingbirds are active during the day (Jones and Davis 1984), while barn owls are nocturnal and only active at night (Smith and Kirk 1982).

If your resource has more than two authors:

Hummingbirds are active during the day (Jones et al. 1984), while barn owls are nocturnal and only active at night (Smith et al. 1982).

If your resource is a website with no author:

Hummingbirds are active during the day (worldwildlifefund.org 2008), while barn owls are nocturnal and only active at night (nature.org 2008).

The Literature-Cited Section: The literature-cited section, also referred to as the bibliography or references, is where you will type out all of the publication information. This is found after the body of the paper has concluded and is usually not included in the overall word count.

To Cite a Peer-Reviewed Journal Article in CSE Name-Year Format:

Last name First Initials. Year of publication. Title of journal article. Name of publishing journal. Volume number (issue number):beginning page number to ending page number.

If there is no issue number, you can place a "1" in the parentheses. Sample citation:

Jones VW, Smith LV, Patel M. 1984. Activity periods of birds. Journal of Bird Research. 34(2):182–189.

To Cite a Book in CSE Name-Year Format:

Last name First Initials. Year of publication. Book title. Location of publication: Name of publisher. Total number of pages.

"pp." is placed at the end to denote total number of pages. Sample citation:

Smith IM, Barlow TK, Jerome MM. 1982. The Life of the Barn Owl. Tucson, AZ: College of the Desert Press. 333pp.

To Cite a Website in CSE Name-Year Format:

Name of organization that created the website. Title of Web page you cited in your paper. Date of Web page publication. Entire URL. [the date you personally accessed the Web page]

You may not find a publication date on a Web-only copy. Just type "date unknown" in lieu of a publication date. Sample citation:

World Health Organization. Malarial mortality in Africa. 3 July 2001. *http://www.who.int/rbm/Presentations.* [accessed 2003 May 5]

* Note: do not put actual boxes around your citations in your final paper or presentation. They are here just to separate the examples for easy reference.

6 Editing

"Edit" is not a dirty, four-letter word. Every good writer edits his or her drafts multiple times. Ernest Hemingway once said, "Write drunk, edit sober." Although I am in no way encouraging you to get drunk to write your paper, this quote is a good reminder that everyone edits—even the greats. Many times, students feel that after that first draft, they're done and ready to submit. If you want a score that reflects the dedication to success I know you have, then editing is a must.

Once you receive your writing assignment, make a schedule for how many pages you will have written by certain dates. For example, if your final ten-page assignment is due one month from now, commit to having at least three pages completed each week. This will leave you with a solid week to edit before the deadline. You should leave yourself enough time before the due date to edit your paper at least twice.

A Self-Editing

Self-editing can be challenging during later drafts, but is the best way to go for the first few. My best recommendation is to read your drafts aloud to yourself. Go somewhere quiet where you feel comfortable talking out loud to yourself. Once you have a good working draft, read the entire paper out loud. Physically speaking the words on the page—and not just hearing them in your mind—helps immensely. It may sound odd, but when we have been working on a draft for a long time, it can be hard to pick out mistakes. When we hear the words, it's much easier to find grammatical inaccuracies and errors in logic. Run-on and awkward sentences show themselves clearly when we speak the words aloud. If you are struggling to say a single sentence or you must take multiple breaths to get that sentence out, it's time to adjust.

B Outside Editing Assistance

Editing advanced stages of your own work can be extremely difficult. Errors in grammar, syntax, tone, or logic all seem to disappear in the haziness of a draft that has been worked, reworked, and reworked again. Statements or ideas that make perfect sense to us may not translate as well to others. This is why it is vitally important to get feedback from an outside source. Although I highly encourage you to get as much outside feedback as possible, remember: not all feedback is of the highest quality. Take into consideration the source of the feedback. What qualifies this source to support your success, and what vested interest, if any, do they have in your opinion or feelings? Lastly, do not take constructive criticism personally. Successful people are always looking for ways to improve. Getting feedback and addressing quality critiques will help you in your writing and your ability to incorporate feedback into your personal routine.

Here are some ideas:

Trade with a Classmate: A fantastic (and free) option is to trade papers with a fellow classmate, preferably an equally dedicated student. This student will have firsthand knowledge of the assignment and an understanding of what the instructor is looking for. Also, a colleague will generally be more comfortable giving you constructive feedback than close friends. A close friend or loved one will usually hesitate to give you any negative feedback because they have a vested interest in your shared relationship and happiness. A colleague you're not personally close to will not feel that strong desire to please you, which will allow you to get better feedback.

Editing the work of others is also a great way to improve your own writing. Whether or not the paper you are reviewing is good, the act and process of editing another's work help you to hone your critiquing skills. If English is not your primary language, please make sure you do not pick up any bad habits from editing others' work. Edit with professionalism and diligence to excellence, but do not let it negatively affect the quality of your writing.

Create an Online Writing Space with Fellow Classmates: Online collaboration can be an excellent tool to get and give quality feedback. It can also be a complete disaster. The best way to utilize an online community is among classmates or already vetted reviewers. E-mail and Google Docs can be ways of sharing, but also look into emerging online communities created especially for college students like Study Room at https://www. getstudyroom.com

Make an Appointment at Your Campus Writing Center: Most campuses have a writing center to support students with their communication assignments. If you are unsure, search your campus website or ask your instructor. A writing center is generally staffed with senior students who are there to help you with grammar, structure, and syntax. Be advised that they most likely will not have the expertise to help you with the science content of your paper, but they definitely can help you become a better writer. Most writing centers are booked far in advance, especially for large campuses. Book your appointment early and give yourself enough time before the due date to effectively use the feedback you receive. Booking an early appointment will also help to give you a personal deadline to meet.

When you arrive to your session, make sure to bring your paper in hard copy and electronic form. Also, bring a copy of the writing assignment with you so that the writing center specialist will know the needs of your particular assignment.

Hire a Professional Editor: Hiring a professional editor does not mean you are hiring someone to write your paper. Professional editing services can be a great resource, if you are willing to

> Grammar: The system and structure of language
> Syntax: The specific arrangement of words in a language that results in well-formed sentences
> Tone: The voice, feeling, or spirit of a creative work
> Logic: Reason or sound judgment

pay the fee. Kibin.com provides feedback quickly and handles everything online. Again, they will not write or rewrite the paper for you; to do so would be unethical and an act of plagiarism. They will, however, have a trained editor review your paper for grammar, tone, and logic. They will not review the accuracy of your scientific content.

Attend Office Hours to Review Your Paper with Your Instructor: Not all instructors will provide an opportunity for feedback on your paper before it's submitted for credit, but many do. The best way to find out is to attend office hours. Writing is a complex process, and it's best to flesh out ideas together, in person. E-mail is great for quick questions, but not for getting quality writing feedback from your instructor. Do not just send off a draft and expect the instructor to edit it for you, unless this was a previously agreed-upon arrangement. Since your instructor will be grading your paper, he or she will be the best source of feedback.

PRESENT!

Oral Communication in Science

7 Presentation Best Practices

The best presentations are like conversations between the presenter and the audience. Whether it's a conversation with three people or 300, if the audience feels comfortable with the presenter, it's better for all involved.

Let's start with some specific ways you can turn the most awkward 20 minutes of your life into the best. Okay, maybe not the best—but at least less horrific than you're imagining right now!

A Preparation

PRACTICE

What's the best way to ace your presentation? Practice! Although there are a few—and I do mean a small few—who can fly into a presentation cold and still wow the audience, most of us mere mortals need to practice before showtime. Actors rehearse before taking the stage, athletes train before a game, and great presenters practice before a presentation. Business leaders practice, politicians practice, doctors practice, so why should you be any different?

Practicing gives you confidence and gives you the opportunity to find out what doesn't work. Imagine if your professor, the CEO of your company, or your president walked into a presentation without having practiced first. You would feel cheated, and your trust in this leader would be significantly depleted.

Prepare for future success now by stretching your presentation muscles in the generally safe space of the classroom.

PRACTICE TIPS

Whether your presentation is given individually or with a group, here are some tips to making your preparation time as efficient as possible:

- If you're using a slide deck, practice the entire deck. Even if it's not finished and some slides are only placeholders, practicing aloud will help with organization and flow.

- Set a stopwatch, and don't stop until you're finished. This will give you a realistic view of how long or how short your presentation is.

- Set a start and end time for your rehearsal. This is especially important for group work, but is also helpful for solo practice. Committing to a set time frame helps to maintain focus and gives the mind a finish line to race for. A quality practice session should produce quality work, and the mind wants to know when it's time to get off work.

- If you're working in a group, have a general agenda for your time together. What are some goals you would like to accomplish during the one hour between classes where you all could meet? Working in groups is challenging during college for many reasons, including scheduling. Having an agenda keeps everyone on task and can move the group back to topic quickly. It can be simple and agreed upon before the actual in-person meeting.

- Once you have practiced at least once, see if you can rehearse in the physical space of your final presentation. Every campus is different, but most have times where the room will be empty. By getting comfortable in the space, finding the plugs and the buttons, knowing what it looks and feels like from the other side of the desk, will all help you feel more confident on presentation day. *Please note*: Some campuses ask that you request permission to be in a classroom during non–class hours. Some campuses have an open-door policy for their classroom space when formal class is not in session. Please make sure to check the campus policies around the usage of instructional space.

PRACTICING FOR A SOLO PRESENTATION

Even if you are presenting by yourself, it's important to practice your presentation before showtime. Having a live audience for your practice presentation is the best option. Using fellow classmates is perfect because you can take turns being each other's audience. Second best would be an extremely kind friend or loved one. But be prepared to bribe them with pizza as a reward for enduring 20 minutes of a graphic-heavy discussion on the prevalence of flesh-eating diseases throughout the tropics.

If you don't have any classmates or friends willing to be your audience, it may be time for new friends. Then, after that, it's time to use your digital camera for more than taking selfies. Most digital cameras have a video recorder feature. Using a tripod or tall dresser, position the camera to record you from at least the waist up. You also want to ensure that your voice can be picked up on the internal microphone. If you don't have a digital camera, your camera phone or laptop may be second best. Following the tips above for practice, give your entire presentation without stopping. Even if you mess up, keep it as close to the real thing as possible. Once you're finished, here is the hardest part: play it back and actually watch it!

I know, I know—nails on a chalkboard have nothing on the excruciating torture that is listening to your own voice and watching yourself on camera. Try not to be hypercritical of yourself, and keep in mind how awesome you are for your dedication to becoming a more successful communicator. Preparing now will make presentation day infinitely easier.

Reviewing a video recording of your presentation is the best way to analyze your performance and start to devote some focus on ways you can improve. However, don't obsess over your performance. You don't want perfection; you want to come across polished and confident. Our own idiosyncrasies make us real; don't lose the vulnerability that your audience will love by becoming perfectly mechanical.

GRADING YOUR OWN PRESENTATION

If your instructor provided you with a grading rubric before presentation day, use this to practice grade your own presentation. If you're working with a live audience, have them use the rubric during your presentation to mark where they feel your performance falls. If you are reviewing your solo video, print out the rubric and have it in front of you while you watch the video playback.

Here are a few common behaviors to look for that you may want to modify in order to deliver the best presentation possible:

- How many times do you say "um," "so," "like," or other filler sounds during times of thought? Can you try and replace the "ums" with silence?

- What are your hands doing? Are you gesticulating too much, thus looking like a chicken trying to take flight? Or are you white-knuckling the podium?

- Are you using your space? Do you fill your space with your body to help create presence and keep the audience engaged? Or do you stand frozen in fear like a deer in the headlights of a speeding car?

Slide Deck: A group of presentation slides
Rubric: A written grading tool that communicates specific expectations for academic assignments
Gesticulating: Using physical gestures to communicate

Aspects of an Engaging Presentation

VOICE

It's time to get real about your voice. Do you naturally talk like a quiet mouse, squawking jay, or subdued sloth? Does it get worse when you're nervous? Then it's time to work on that. If you honestly don't know, ask an independent opinion from a trusted source. More often than not, we think we sound a lot worse than we actually do. Sometimes we think we sound better than we actually do; either way, it is vital to get the truth.

Find someone who is more of a colleague than a friend, and definitely don't ask a relative: you want someone who is impartial and doesn't care about your feelings. A classmate is great; you can each trade reviews of one another's voice strength.

There is no perfect definition of the perfect speaking voice. Yes, there is too fast or too slow. There is too quiet or too loud. But trying to define the best attributes of an amazing speaking voice is like trying to define the requirements for a great painter or iconic singer. You know it when you see it—and you sure know when you don't. With your voice partner, openly discuss perceptions of each other's performance. Watch videos of professional speakers (think Ted Talks, ted.com), and compare notes on what most attracted you to their speaking style.

GESTICULATION (NO, THAT'S NOT A DIRTY WORD)

Gesticulation means emphasizing speech with one's hands. Not in a formal language sense like sign language, but a natural habit of reinforcing what is said with hand movements: sweeping motions signifying vastness, finger explosions for really exciting points, etc. Gesticulation can be a wonderful way to illustrate and reinforce your points. It can animate your presentation and draw in your audience. It can also go horribly wrong by distracting your audience from the presentation's content.

So what to do? Well, if you're already a natural gesticulator, meaning when you're calm and relaxed you still talk with your hands, keep doing it. Not sure?

Ask your friends and family; they will know. Think about how you can use your gesticulation habits to further emphasize main points, but don't really worry too much about it. If you don't already use gestures when speaking, my recommendation is don't start now. Again, it can be an advanced way to emphasize your points, but at this stage in your presentation career, it's better to execute a simple, but effective, delivery.

EYE CONTACT (BUT IF I LOOK IN THEIR EYES, I'LL LAUGH ... OR CRY ...)

Looking directly in the eye at your peers or your professor during a presentation is the perfect way to completely forget everything you practiced. It can be too unnerving to look someone directly in the eye, especially if they aren't so excited about your presentation. It can break your concentration and your confidence. But eye contact is essential for making audiences feel like you're having a conversation with just them—and only them.

This is where the cliché "fake it 'til you make it" fits nicely: look at the tops of their heads instead of their eyes. You will most likely be at least five feet away from your audience, which gives you a bit of wiggle room when it comes to perspective. Your audience will not be able to tell if you're looking them directly in the eye or just a little higher.

Additionally, you want to include everyone in the room. You want to gently move your gaze to all parts of the room, from left to right to center. Looking like you're watching a tennis match is extremely uncomfortable for your audience. Delicately move your eye contact across all parts of the room, paying special attention to avoid looking at the ceiling or your feet.

EFFECTIVELY USING YOUR SPACE

Warning: This is an advanced maneuver. If your primary goal is to deliver a presentation without vomiting in front of the class, then save this section for later. But hey, maybe you're already comfortable on the stage and are ready to turn things up a notch.

Using your space is not standing in one spot. It is physically moving your body to different parts of your stage. When you move, so does the focus point of your audience's attention. It's using your hands to point to a graph on your slides, or walking out from behind the podium or desk to get closer (but not too close!) to your audience. When you effectively use your physical space during a presentation, it makes the content more vibrant and makes you seem more interesting, thus keeping your audience more engaged.

AUDIENCE PARTICIPATION

An effective, but tricky, way to include your audience in your presentation is to elicit their participation. Some of the more traditional ways would be:

- Question-and-answer

- Small group breakout

- Pre-chosen group breakout

You may also include handouts or prizes here. As with all presentations, there is a delicate balance between a performance being memorable for all the right reasons—or all the wrong ones. Two examples taken from my classes:

- A group of students presenting on sexually transmitted diseases to their university nonmajors course informed students that they could get up to four free condoms per day from the student health center. Before their presentation, they arranged with the student health center to secure around 150 free condoms for their audience. After introducing this free service, the group handed out the condoms to everyone in the audience. Every student was definitely awake and paying attention at that point. Winning!

- One group wanted to engage the classroom by having a question-and-answer session. When someone would correctly answer a question, they would get a purple Hershey's Kiss thrown their way. You know where this story is going. Yes, you guessed it—a Hershey's Kiss right in the eye. Not winning.

The moral of these stories is to be bold and be creative. Be aware of what can go wrong, and try to anticipate it before it actually occurs.

PROPS

As an undergrad, I once gave a presentation on marine mammals dressed as a sailor. Yes, that really happened. The students thought it was great; my professor ... not so much. But don't just learn from my mistakes; learn from some of my students' successes! One group, presenting against drilling in the Arctic National Wildlife Refuge, brought in homemade masks, anti–oil industry signs, and 3D diagrams. In addition to a well-thought-out and practiced presentation, the props that the group made were creative and memorable. They helped to reinforce their assigned position and left the audience wanting more. Unlike my sailor costume, they set the stage for their presentation before beginning to speak. They left everyone in the room with a message to remember.

PERSONAL STORIES

Humans love a good story. It's so ingrained in our culture that we are gladly willing to sacrifice our time and money to experience it. A good story, from the heart or from personal experience, can skyrocket your presentation. That being said: don't make it up. It doesn't matter if your Oscar-worthy performance would bring the class to tears of joy and/or sorrow. If it is a lie, don't do it. One of science's greatest goals is honesty. Does that mean every scientist who has ever lived has been the epitome of virtue? No, of course not. But the goal is to try. Science is about the truthful pursuit of what is or what could be. Never include a story that is not true or pass one off as yours when in fact it's not.

HUMOR

Using humor is a fantastic way to engage your audience. But on the academic stage, you must be careful. Unlike a comedy club, your audience members will not be under the influence of alcohol (hopefully), and will most likely find politically incorrect content to be offensive. Finding that delicate balance between funny and abusive is absolutely possible—you just need to work at it. Be mindful of the way your humor may be perceived by your peers and your professor. In an undergraduate nonmajors course, the two main goals are to get your message across and get an excellent grade. Therefore, it is best to be cautious with humor in an academic setting.

A lot of humor is rooted in making others feel like less: less than adequate, less than acceptable. It is a well-crafted jest that allows everyone to laugh equally. Use humor whenever possible, but make sure you think of your audience. Think of the millions of stories and experiences your audience brings into the room that will color their interpretation of your humor. Making fun of plants is still funny, and rock jokes are good, too.

POP CULTURE REFERENCES

Making your presentation relevant to your audience is one of the best ways to engage them. It's natural to turn off and tune out when we think what is being said doesn't apply to us. We are constantly bombarded with information, messages, and all the things we "should" be doing. One way to bring your audience back to your message is to talk about things they can relate to. Enter pop culture. I will by no means illustrate current examples of pop culture here, since popular culture—aka pop culture—changes daily in today's lightning-fast, social media–driven environment. However, by definition, pop culture is the collective relevancy of media icons, videos, films, scandalous stories, and hot topics of the day. If you can draw a clear—and I mean crystal clear—connection between your topic and a pop culture reference, use it. If it is less than 100 percent related, leave it out. Just like other methods of audience engagement, used correctly, pop culture references can wow a crowd. Used inappropriately, they can offend not only your peers, but also your professor, who will be directly responsible for critiquing your performance.

Mark Kreuiter

Tables and chairs started scraping, bashing, and bouncing against and off each other loudly, resembling a rowdy bunch of prisoners getting out of hand in a mess hall, while my Cuisine and Culture Professor looked unamused. I've always been anxious about public speaking, so to calm my nerves, I decided to make my Greek cooking demonstration and speech on Greece interactive in the hopes of diverting attention away from me. Toward the end of my presenation, I called up five volunteers to perform a classical Greek dance with me in the front of our realtively small classroom. As Greek music blared from my mobile phone, we all really got into the dance, with movements becoming more frantic and turbulent in pace with the accelerating music.

Up until this ending, I had done what I could to deliver a strong presentation. I fully understood my material so I didn't need to rely on notes. I stood tall and did my best to focus on someone new in the audience for each sentence, so he or she would feel personally included. The only aspects I didn't account for were the physical constraints of the space and the specific attributes my Professor, arguably the most important audience member in terms of my grade, was looking for.

In the end, I was proud of my creativity, my ability to take risks, and the bonding it created with my classmates. In the future, I would only slightly suggest getting approval before hand, but at least that was one presentation my Professor will never forget!

Mark Kreuiter
www.digitfields.com
University of South Africa
South Africa

8 Creating Engaging Presentation Materials

Communicating effectively—whether through writing, visuals, or oration—is the key to success. No matter what the field, good communicators are more successful than inadequate communicators. This especially holds true in highly competitive fields, such as business, technology, and science. In this section, we will discuss how to make your digital presentation, sometimes referred to as a slide deck, engaging and memorable for all the right reasons.

A slide deck, whether you are using PowerPoint, Keynote, or Google Slides, should complement your oral presentation, not distract from it. Slides are meant to enhance the brilliant words you're sharing with your audience, not act as a giant glowing script that you read from, because you're unprepared.

Think back to presentations you have seen in the past. This could be from a fellow student, a teacher, or even a video of a presentation like Amanda Palmer's TED Talk or Al Gore in *An Inconvenient Truth*. Then check off what you loved and what you didn't below.

CHECKLIST

Made me Smile / Made me Cringe

The presenter:

Read directly from the slides/note cards _____ _____

Looked down at his/her feet or only at
his/her notes _____ _____

Spoke softly _____ _____

Spoke too fast _____ _____

Spoke too slow _____ _____

Used hand gestures to emphasize points _____ _____

Made direct eye contact with you and the audience _____ _____

Equally moved his/her gaze across everyone in
the room _____ _____

Stood still behind a podium or desk _____ _____

Moved to different parts of the stage/classroom _____ _____

Turned his/her back to you and the audience _____ _____

Spoke in a monotone voice _____ _____

The Slide Deck:

Each slide was filled with lines of text _____ _____

The font size was very small _____ _____

Every slide was in a different font, different
colors, and different backgrounds _____ _____

There was minimal text on each slide _____ _____

There were many pictures, graphs, and charts _____ _____

The text color was in sharp contrast to the
slide background _____ _____

Building Your Slide Deck

No matter which program you choose to use, there are some basic elements that help to create a powerful slide deck. Start by looking at your checklist. Which aspects of past presentations made you enjoy them more and which aspects made them hard to watch. You can't always control what happens when you're on stage, but you can control how much you prepare. So even if you vomit on the podium in front of the class, at least your slide deck will look fantastic!

Let's start with the basics and build from there.

FONT

Unless your instructor has made it clear which font to use, there is no industry standard font. There are, however, good fonts to use and not so good fonts. Whichever font you choose, it must be clear enough to read from the back of the room. If your font is too messy or abstract, it will be indecipherable. In an academic presentation, all text needs to be clearly read from any seat in the room. Stay away from fonts that are animated. Unless there is a very specific reason, fonts in an academic presentation should not sparkle, glow, or have flames shooting out from the capital letters.

Look for a font with clean lines that is as easy to read at 12 point as it is at 55 point. Bolded text should be used sparingly, and shadowing should be used only when absolutely necessary. Make your presentation visually engaging with directly related graphics, not wild fonts. The most commonly used fonts are Arial, Times, Times New Roman, and Calibri. Here are a few other great choices that are widely available in slide deck programs and all are in 14-point font for reference.

Biology (Font name: Iowan Old Style)
Physics (Font Name: Century Gothic)
Chemistry (Font Name: Georgia)
Geology (Font Name: Osaka)
Environmental Sciences (Font Name: Euphemia UCAS)

Here are some poor choices in font style that you will want to stay away from. Notice how they are hard to read, even up close in this book. These fonts would make it difficult for your audience to read along while you're speaking. Each font here is also in 14 point.

Biology (Font name: Mistral)
Physics (Font Name: Matura MT Script Capitals)
Chemistry (Font Name: Playbill)
Geology (Font Name: Xingkai SC Light)
Environmental Sciences (Font Name: Chalkduster)

FONT SIZE

Depending on the room size and style of font, most of your fonts should be between 35 and 55 point in size. Again, you want every person in the room to clearly see the textual content of your slide deck. If it is too small, the purpose of having a slide deck is lost. You should only use a smaller font when appropriate, like when giving credit for a graphic in a byline, using slide numbers, or using master titles on each slide.

FONT COLOR

You want your font color to be in distinct contrast to your slide background. Choosing a color too similar to your background will make the text incredibly hard to see. If you're using a dark background color (which actually saves a tiny bit of electricity), then you want to use a white or very light colored font. Conversely, use a dark font color for light backgrounds. Don't be afraid of using color here. Feel free to explore beyond the black and white, just make sure that the font colors contrast your background, making them easier to read.

Less is best when thinking about how much text you should place on each slide. Unless you are including a quote of epic proportions that you absolutely must read to the class famously uttered by someone relevant to your topic, never put paragraphs of text on your slide. Think bullet points and keywords over full sentences. The point of having a presenter is to hear what the presenter has to say. This is one of the biggest challenges I see for students. If you have an enormous amount of text on your slides, you will end up reading it directly to the audience. Reading slides verbatim is terribly boring for your audience and shows your instructor that you did not practice or prepare, thus hurting your grade.

A slide deck is meant to enhance your presentation—not replace practice and preparation. The slide text should help outline your points and reinforce what you're saying. Focus on bullet points, keywords, data, statistics, and very short sentences.

PRESENTATION NAME

SOME TITLE HERE

- Lorem ipsum dolor sit amet consectetuer odio non tellus natoque accumsan.
- Sed hac enim Lorem tempus tortor
- Ejusto eget scelerisque sed morbi.
- Senectus urna Vestibulum tincidunt turpis

PRESENTATION NAME

SOME TITLE HERE

- Lorem ipsum dolor sit amet consectetuer odio non tellus natoque accumsan. Sed hac enim Lorem E usto e et sceleris ue sed morbi i sum dolor Lorem ipsum dolor sit amet consectetuer odio non tellus natoque accumsan
- Sed hac enim Lorem tempus tortor non tellus natoque accumsan. Sed hac enim E usto e et sceleris ue sed morbi i sum dolor Lorem ipsum dolor sit amet consectetuer odio non tellus natoque accumsan consectetuer odio E usto e et sceleris ue sed morbi i sum dolor Lorem ipsum dolor sit amet consectetuer odio non tellus natoque accumsan consectetuer odio

SLIDE BACKGROUNDS AND THEME COLORS

The software selections reviewed below will have a varying degree of slide background design and color options. Choose wisely.

Ask yourself these questions when selecting a slide background and color scheme:

- Is it visually pleasing but not distracting?

- Is there enough space to put words and graphics on the slide without overlaying on the borders?

- Is the entire background an enlarged photograph or detailed graphic? If you answer yes to this question, I highly suggest you replace the background. Using photographs or detailed graphics as background is extremely distracting, making it impossible to properly line up other images and creating a challenging reading environment. Pictures are essential to a great presentation; just make them the furniture not the wallpaper.

- Are the colors easily contrastable? Pastels (light colors) with other pastels do not work well for projection. Use either two bold main colors or a pastel with a very dark color.

FRANKSTEIN'S MONSTER SLIDE DECK

Frankenstein's monster, a fictional character that was created by a scientist with less than exemplary ethics, was created using parts from multiple corpses. With digits from Danny, ears from Elaine, and lips from Laura, the monster looked quite a mess! Don't make your presentation look like Frankenstein's monster by taking slides created and designed by multiple people and simply pasting them together without any additional formatting.

When working in a group where each member is responsible for creating slides, designate one person to be the Master Formatter. Everyone in the group will send their finalized slides to this person without any visual formatting—which would mean a blank white background and plain black text. The Master Formatter will then be responsible for playing surgeon and putting all the pieces together. He/she will also be responsible for ensuring all slides have the same background, the same font, and the same font size for different attributes (i.e., titles, subtitles, and content text). The look and feel of the entire presentation needs to match—this is non-negotiable in any type of situation. If you don't take this extra step, your slides will show your lack of preparation, and it will most likely negatively impact your grade.

The Master Formatter of the group should also be responsible for any fancy formatting that needs to be done, like animations, video embedding, and transitions. You and your team should meet to decide on the specific design elements (background theme, colors, fancy formatting, etc.), but ultimately, it should be just one person taking care of all the final touches.

I hope it goes without saying that your elected Master Formatter also needs to be a trustworthy student. The Master Formatter will be responsible for getting all of the group's hard work to the finish line. When you meet, the entire group should agree on a date—far in advance the official presentation day—to have the Master Formatter email or post the finished version to the group. Every person in the group should have a copy of the final version to ensure that, even if the Master Formatter misses the presentation, the show can still go on.

GRAPHICS

You need graphics. This is a must for a visually engaging presentation. People connect through visuals, especially now. We SnapChat, Instagram, and Tweet pictures as a way of communicating. Many times, we can create entire conversations without using a single word. Graphics help engage your audience and illuminate your points. Using graphics is a perfect way to help give your audience a visual accompaniment to your verbal greatness.

Some of your graphics will only serve as decoration, and some of your graphics will be instrumental to what you're saying. For those graphics that are really important, like a graph of your data, a diagram of a molecule, or an illustration of sediment layers; turn one quarter to the screen and use your body to physically include the graphic. You can do this using a laser pointer, the mouse on the computer, or motioning with your hands. Only turn one quarter though, you never want to have your back to the audience.

STILL GRAPHICS

Still graphics do not move. These are pictures, charts, graphs, maps, tables, and any other images that are not animated. Generally, for undergraduates, you will be using graphics that someone else has created. Most undergraduates will lift their images from websites that they do not own, and getting copyright release for a class project is pretty unfeasible. At minimum, you should be giving the creator credit as a small byline next to the image. More information on copyright law and how it affects students can be found in Chapter 12.

Milkweed butterfly – © Ansonde

When choosing a graphic for your presentation, make sure you have a high enough resolution so that it doesn't look pixelated when projected. Most search engines will allow you to sort graphics by file size. Go for the largest file size available so that it will display clean and clear to your audience.

Line up your graphics or display them in some type of logical pattern.

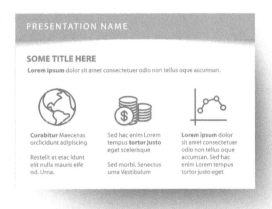

Stay away from animated Graphics Interchange Format files, otherwise known as GIFs. These are those short animated images. They can be fun to post to social media or over instant message, but are extremely distracting in an academic presentation.

FANCY FORMATTING: EMBEDDING VIDEOS AND ANIMATIONS INTO YOUR SLIDE DECK

Using videos in your presentation, as long as it's allowed, is a great way to engage your audience. Depending on the total length of your presentation, having an animation or short video can help to illustrate your concepts and keep your audience awake. Here are some key elements to remember:

- Make sure it is 100% relevant: When choosing an animation or video, only include those that can be easily connected to your main topic points. If you have to ask yourself, "I wonder if they will get how this is related?" then the answer is probably no.

- Make it short and sweet. Timing is everything: In life and in presentations. Make your chosen video/animation "short" relative to your total presentation time. If your total presentation time is ten to twenty minutes, each video should be no more than 30 to 60 seconds. A two-minute video/animation is really the upper limit to play during a presentation 20 minutes and under. It is also a good rule of thumb to include no more than one video/animation every 5 minutes; otherwise, it makes the presentation seem like you are just sharing your video playlist with the audience.

- Make it professional, and remember your audience. Even though you may think your analogy or example may be a perfect fit, your audience—and, more importantly, your professor—may not. Always err on the side of being more professional and considerate. Since you never know who will be in your audience and in an academic setting, it's better to keep it professional. Using humor, pop culture, anecdotes, and personal stories are all amazing ways to get your audience excited about your topic—just make sure to take a step back and really ask yourself if the content is appropriate for the class and what are the chances your audience will be offended by its content.

B Content

What is the point of your presentation? In order for your presentation to make sense to your audience, you must be confident in knowing what you want to say. Start by outlining the major points you want to cover in your presentation before you begin creating it.

What do you want your audience to feel after your presentation?

Boredom and relief that it's over are probably not top on your list. However, each presentation is an interaction with you and your audience. How do you want your audience to feel after you have finished? Do you want them to feel empowered, because you just gave a talk on heroes of the environmental movement? Do you want them to feel more informed, because you discussed the most common sexually transmitted diseases for college students and where they could get free condoms on campus?

Trying to figure this part out can be hard sometimes, especially with content that lacks human traits or doesn't immediately evoke an emotional response. Sometimes, it just can't be done. If you were presenting an equation proof, the highest emotional reaction you could hope for would be comprehension. Even if you don't think eliciting an emotional response from your audience applies to your particular project, just keep it in mind: humans, even scientists, are emotional creatures. We learn and remember better when we have an emotional response. We can recite songs that make us cry or lines from a movie that made us laugh, but we can't remember anything from that dry science book our instructor forced us to read for an exam.

What do you want them to remember? The Big 3 Method

This one may be a little easier to answer. Ask yourself, "what is the point of your presentation?" Well, besides scoring a good grade. Since most class presentations will be between 5 and 20 minutes, try and narrow your focus down to 3 big ideas or messages you want your audience to remember after you're done. Then, build your content around those Big 3.

When I begin to build a presentation, I start by answering these questions:

- What topic have I been given/chosen to speak on?

- If I could only have people remember **one** thing from my presentation, what would it be?

- If I could only have people remember **two** things from my presentation, what would they be?

- If I could only have people remember **three** things from my presentation, what would they be?

- How do I want my audience to feel after the presentation (as applicable)?

Let's say I'm doing a presentation about climate change, which is an enormous topic. I only have ten minutes to present, so I need to do some serious focusing. This is where the Big 3 Method comes in. Since I know my audience will be made primarily of college students, I want to talk about how climate change is relative to them. I focus on creating an empowering presentation on three areas where college students can take immediate action against climate change. I choose: Decreasing reliance on a meat-based diet to help decrease global methane emissions, drinking from only reusable water bottles to reduce petroleum consumption, and purchasing sustainable rainforest products to help protect the world's largest carbon-sequestering terrestrial ecosystem as my Big 3. Now, I can begin to build my presentation around supporting these three main topics.

After I have set these intentions, I go back and see how feasible my Big 3 are given how much time I've been allotted. Don't be afraid to change and edit your Big 3 as needed.

Climate Change: Long-term change in Earth's climate, which currently involves an increase in average global temperatures
Carbon Sequestering: The capturing and storage of carbon atoms through physical or biological processes

C Projecting Your Presentation On The Big Day

USING YOUR OWN TECHNOLOGY

Run through the entire presentation at least once in presentation mode on your computer to ensure everything works. This is especially true if you have included pieces from multiple partners that were created on different machines. There may be software updates you need to install, and you definitely don't want to do this during your performance.

How will you connect your computer to the projector in the room? Unless you're bringing your own projector with you, make sure to check out the projector you will use well in advance of your presentation day. No matter what type of technology you're trying to use to display your presentation, make sure that it can connect to a projector. Most tablets do not have this ability, although this is beginning to change. Most full size laptops do, but not all. Many mini-laptops and most netbooks also lack the ability to connect to a projector.

If you have a newer Mac laptop you will most likely need a Mini DisplayPort to VGA Adapter or HDMI cable, since there is no built-in VGA adapter. Do not assume the instructor will have these cables for you. Depending on the projection technology, you may need some other type of physical adapter, although VGA input is the most common, with HDMI becoming more common with newer displays. If you have a PC, check to ensure you have a VGA input available.

Lastly, if you're using sound, double check that the room you're presenting in has speakers. Playing a video on the overhead screen while the audio plays from your tiny laptop speakers is going to make for an unenjoyable viewing experience.

The best thing to do is do a short test run in the actual room you will be presenting in. Even if it's just a few minutes after class, ask your instructor to show you how the buttons work and how to adjust sound and plug in your system. Do a quick run through of your presentation file to verify everything looks and sounds fantastic.

USING THE TECHNOLOGY PROVIDED IN THE ROOM/BY THE INSTRUCTOR

If you will be using someone else's technology to project your presentation file, make sure everything you have created will work before presentation day. Checking compatibility before your presentation date will save you a lot of undue stress and frustration. In the following list, we will use *Destination Machine* to mean the technology, that is not your own, that you will be displaying your work to the class. We will call the machine(s) you originally created the work on, *Origination Machine*.

Things you want to check for:

- Are all the software programs you will be using updated on the Destination Machine? Don't forget to check for auxiliary updated software that you may be using, such as Adobe Flash Player (http://get.adobe.com/flashplayer/) or Apple QuickTime (https://www.apple.com/quicktime/).

- Will you be using sound? Ask if the Destination Machine has sound capabilities. If so, ask your instructor to show you how to control volume levels before your presentation day. Being nervous on top of not knowing how to use the technology sets you up for a rough performance.

- Will you be playing video? If so, check that the Destination Machine is able to play the videos you wish to show. Most students will use web videos that require specific software in order to play. If the video player software, such as RealPlayer, is outdated, then your perfectly timed video may not play.

- Will you be using the internet? If so, does the Destination Machine have access and is the browser (such as Firefox, Safari, Chrome or Internet Explorer) up to date? If the browser is outdated, then it could cause major problems with displaying the web content you wish to share.

- Do you have a backup plan in case the internet crashes or is unavailable? Don't always assume that the internet connection will be reliable or even available during your presentation. It most likely will, granted you have verified with your instructor that internet is accessible. However, always have a backup plan. Have your presentation saved on a thumb drive, also known as a flash drive. Even if you are using a Web-based slide deck platform, save a copy that is accessible without internet access. Can you download the video clip you want to use? If so, do it and save it to your flash drive. One last backup option for internet problems is tethering to the internet via a smartphone.

Lastly, always have a backup file easily accessible. Make sure everyone in the group has a copy of the final presentation stored in their email, Google Drive, or other cloud storage, so that it can be easily accessed from the classroom or presentation hall immediately, if needed. Never depend on one person to hold the only copy of the final presentation. If that one person does not show up, then the entire group is in jeopardy. It's better to have a backup and never use it, than to have no backup and fail your presentation.

 ## Slide Deck Presentation Software

Microsoft PowerPoint has been the gold standard of digital slide creation for decades. PowerPoint is a great option, but there are other fantastically free options available to help create your digital presentation. Below are my favorite software choices that are easy to use and functional.

PowerPoint (products.office.com/en-us/student/office-in-education)

- Cost: You must purchase a license, however discount licenses are usually available through your campus bookstore.

- Good for groups: Not really

- Cloud storage: Not unless someone in your group already has a private cloud and is willing to store and share.

- Online collaboration from anywhere: No

- Biggest pro: It's an industry standard, so almost any machine will be able to project a .ppt or .pptx file created by PowerPoint.

- Biggest con: Doing Fancy Formatting in PowerPoint is a lot trickier. It's not impossible, but it definitely is more challenging than Prezi. Also, the background and themes that come standard with PowerPoint are pretty basic and have been used millions of times by other students. You're not going to wow your instructor with the Orbit theme that they've seen every semester since they began teaching that class.

Prezi (prezi.com)

- Cost: Free for Students and Faculty

- Good for groups: Yes

- Cloud storage: Yes

- Online collaboration from anywhere: Yes

- Biggest pro: Sleek design with easy-to-use animation functions

- Biggest con: Sometimes the font can turn out really small in the final product and be hard to read from the back of the room.

Prezi has an EduEnjoy package that provides students and faculty free access via http://prezi.com/pricing-4/edu/. At the time of this printing, this free-for-academics package includes 500 megabytes of storage, secure and private presentations, and premium support.

What makes Prezi really awesome is that it easily allows you to add motion graphics—primarily zooming in and out— to your presentation without having to do any extra animation work. Prezi is structured more like a concept map and less like a linear set of slides. Since Prezi is cloud based—meaning it is stored on Prezi's secure server—you can access it and edit from anywhere. Prezi's free academic package also allows group members to collaborate online on the same presentation.

Google Drive (google.com/drive)

- Cost: Free

- Good for groups: Yes

- Cloud storage: Yes

- Online collaboration from anywhere: Yes

- Biggest pro: Super easy to use

- Biggest con: Incredibly basic formatting options that will leave your presentation looking drab and boring

Google Drive, which houses Google Slides, is a free web-based software that allows multiple users anywhere in the world to not only create, but simultaneously edit a document, presentation, or spreadsheet. Google Drive is stored in secure cloud storage, which allows for it to be shared and accessed by anyone who has the link or permission to do so. You can control who has access to your files within your project.

Google Drive is great for group work and collaboration, but terrible for creating visually engaging slides. If you are going to use Google Drive's presentation function to collaborate with your team online, I highly suggest that you designate one person to do the design once all the content is completed. This one person will be responsible for formatting all of the slide's backgrounds, fonts, and fancy formatting. Please see "Frankenstein's Monster Slide Deck" earlier in this chapter for additional information.

CREATE!

Visual Communication in Science

9 Presentation Video Format

Presentation Video Format

    a. What is a presentation Video?

    b. Software

    c. Hardware

    d. Audio

    e. Quick Tips for Shooting Video

    f. Quick Tips for Computer Screen Recording

 ## What is a Presentation Video?

As online classes become more and more popular, presentation videos are taking the place of traditional lectures and even replacing in-class speeches! Gone will be the days of sweaty palms, a cracking voice, and a dry mouth on presentation day. Well, perhaps not completely. But presentation videos are the newest way students are delivering their presentations in the online classroom environment. Presentation videos are also called lecture captures or screen captures. Whichever name you choose, they are essentially a recording of your computer screen, your voice, and perhaps your face or body.

This type of assignment can also be a wonderful break from the norm for students who are interested in being creative. Most instructors will publish guidelines, but will generally encourage creativity with this type of endeavor. But of course, as always, know your audience and how much creativity your instructor allows. Every instructor is different, some value highly creative works, some do not. Find out exactly where your instructor lies on this spectrum early on before you begin creating your project.

Presentation videos can be captured in a few ways: straight to camera, voiceover, and hybrid.

Straight-to-Camera Style: This is a style utilized in many YouTube videos where the presenter or host of the video is speaking to you by looking directly into the lens of the camera.

Voiceover: A voiceover, or VO, is where the viewer watches moving images of your desktop screen, animation, or slide deck while listening to your voice. This is sometimes referred to as a narration. In this type of video the viewer generally never sees your face/body.

Hybrid: The hybrid style of presentation is where the viewer will see your slide deck, as well as your face and/or body. This can be done in a few ways. You can record yourself standing and presenting your slides in front of a large screen using a video camera or digital video recorder. You can also utilize the advanced features in some software platforms, like Camtasia, to show your slides digitally at the same time, displaying a subset or cutaway of your face or body.

A short note on creativity: Even if you think you are not artistic, be creative anyway. Participating in creative behaviors, such as painting, drawing, writing, dancing, singing, digitally illustrating, sculpting, or video making is an important aspect of living a fulfilling life. Science does not always lend itself to creative projects, but it does so more often than you think. Science and art can and should go together. I encourage you to take every opportunity to put a little more art into the world. Even if you think you can't draw, write, dance, or sing: do it anyway. The world needs more of your creativity and artistic contributions, whatever they may be.

B Software

You can create your digital presentation using a wide range of software options. They differ widely in cost, learning curves for startup, and production quality. Below are reviews of a select group that I have personally used. These are my own opinions, which may differ from yours or other reviewers. The websites and costs were correct as of this printing, but for the most current information, please visit **www.WritePresentCreate.com**. This is not an exhaustive list, but I feel they are the best options available to you now. If you know of other software programs that would be good for the non-major, undergraduate, please let me know at **www.WritePresentCreate. com**

The software programs below listed in order from easy-to-use to more difficult to use, with free options being presented before pay options.

Ease of Use and Sharing Scale: Super Easy > Pretty Easy > Not So Easy

Audio Scale: Excellent > Pretty Good > Acceptable

Editing Scale: Basic > Moderate > Expert

JING

- Cost: Free

- URL: **http://www.techsmith.com/jing.html**

- Pros: The startup time from download to use is minimal. There are very few functions available, so it is extremely user friendly. No complex training or tech expertise is needed. A short URL hyperlink is provided once the recording has finished uploading to the Jing server. The Jing "Sun" is embedded as a quick access tool on your desktop and will appear in your programs.

- Cons: There are no tools to edit your recording. If you fumble, take too long to start up, or sneeze right at the end of a perfect take, there is nothing you can do about it. The file is uploaded to the Jing server, so there is no way to manipulate it after it is recorded.

- Useful for: Short screen captures that are less than 5 minutes with music and/or voice over

- Ease of use: Super easy

- Editing ability: None

- Audio quality: Acceptable

- Ease of sharing with others: Super easy

- Cost: Free

- URL: **https://www.apple.com/quicktime/download/**

- Pros: QuickTime is one of my favorites. It's built into all newer Mac OS X operating systems, so if you're a Mac user, you probably already have it. For basic screen capture, it's great. You can trim the clips and export files in 1080p, 720p, 480p, or smaller. The file type will be a .mov, which is easily uploaded to YouTube or Vimeo. You can select only a portion of your desktop to record or the entire desktop or just record audio.

- Cons: The native ability to edit within QuickTime is extremely limited. However, if you're using QuickTime on a Mac, you most likely already have iMovie. See below for my review on iMovie.

- Useful for: Screen capture of lectures, tutorials, or simple audio recording

- Ease of use: Super easy

- Editing ability: Basic

- Audio quality: Excellent (as long as you are using a high-quality microphone)

- Ease of sharing with others: Super easy

Trim: Removing pieces of a video clip

- Cost: Free

- URL: **http://tinytake.com**

- Pros: It's free and it's for Windows! If you're a PC user and are feeling a little left out of the photo and video software arena, then you'll be happy to find this Windows-only free platform. You can choose to record just a portion of your screen or you can record your entire desktop. You can also record video using your webcam. Video or screen capture can be recorded up to 120 minutes. You can publish and get a shareable link once you are finished.

- Cons: Limited editing capability

- Useful for: Screen captures, presentation videos that need little editing, online collaboration, and tutorials

- Ease of use: Super easy

- Editing ability: Basic

- Audio quality: Pretty good

- Ease of sharing with others: Pretty easy

Export: Saving your file in a format usable by other programs

- Cost: Free with the purchase of an Apple laptop or desktop with Mac OS X

- URL: **https://www.apple.com/mac/imovie/**

- Pros: iMovie has a lot of editing power and comes with very little need for professional training. The user interface is specifically designed to be intuitive so that anyone can upload, edit, and share videos. Although it is not as powerful as Final Cut Pro, for most video projects outside of Hollywood, Cannes, or film school, iMovie is a great option.

- Cons: It will usually increase the file size from your original, raw video files. It will export as an .mp4 which can then be uploaded to YouTube or Vimeo, but you do not have any other choice of file type to export.

- Useful for: Any basic or enhanced video editing. You can bring together as many different clips as you wish and add basic effects, text, trim, music, and a host of other modifications easily.

- Ease of use: Pretty easy

- Editing ability: Moderate

- Audio quality: Excellent (as long as you are using a high-quality microphone). If you start with quality audio input, you will have quality audio output. There are some basic audio tools within iMovie to help enhance poor audio, but don't rely on it to make your poor audio recordings great.

- Ease of sharing with others: Super easy

> **User Interface:** The design elements of an electronic device that a human directly interacts with

SCREENCAST-O-MATIC

- Cost: Recordings that are 15 minutes or less are free. In order to record over 15 minutes, remove ads, and add basic editing capabilities, you will need to purchase their "Pro" package at $15 per year.

- URL: **http://www.screencast-o-matic.com**

- Pros: You can get 15 minutes of web-based screen capture recording for free without having to install any software on your computer.

- Cons: If you go the free route, you will have ads, which will distract from your final product.

- Useful for: Screen or video capture that can be done straight through without stopping, and you won't need any fancy editing.

- Ease of use: Super easy

- Editing ability: Zero for the free version; very basic for the paid version

- Audio quality: Acceptable for short projects not meant for high volume or large-room viewing.

- Ease of sharing with others: Super easy

- Cost: Free trial for 15 days and educational pricing is available at **http://www.techsmith.com/volume-license-education.html**

- URL: **http://www.techsmith.com/snagit.html**

- Pros: Extremely quick and easy with excellent basic editing and sharing abilities. There is also a smartphone app available for Android and iOS devices via TechSmith Fuse at **http://www.techsmith.com/fuse.html**.

- Cons: The cost and lack of full-scale editing ability.

- Useful for: Capturing and editing screenshots and recording your desktop screen with voiceover.

- Ease of use: Super easy

- Editing ability: Moderate—You are able to trim clips, create pre-designed steps to take your viewers through a preconceived logical order, and employ basic special effects.

- Audio quality: Excellent (as long as you are using a high-quality microphone)

- Ease of sharing with others: Super easy via file download or Google Drive

CAMTASIA FOR MAC AND CAMTASIA STUDIO FOR WINDOWS

Although there are two different versions available, they are extremely similar, depending on your operating system. There are a few slight variations, but nothing extraordinarily worth mentioning here, other than cost.

- Camtasia for Mac Cost: Free trial available. Full version direct cost is $99 with education pricing available for $75.

- Camtasia Studio for Windows Cost: Direct cost is $299 with education pricing available for $179.

- URL: http://www.techsmith.com/camtasia.html

- Pros: Extremely powerful. Everything you would want to do for screen capturing you can do in Camtasia. Also, there is excellent technical support provided to get users ready to create videos as quickly as possible.

- Cons: Camtasia has a slightly more difficult learning curve than many of the other web-based platforms discussed. More power with more flexibility and control generally means there is more to learn. That being said, Camtasia is relative easy to navigate with a little time and attention. Lastly, the cost can be prohibitive to some students.

- Useful for: Creating and detailed level editing of screen capture videos, audio recordings, and video lectures (straight to camera, hybrid, or voiceover styles)

- Ease of use: Not So Easy

- Editing ability: Expert

- Audio quality: Excellent (as long as you are using a high-quality microphone)

- Ease of sharing with others: Super easy. File downloads or share straight to Screencast.com, YouTube, or Google Drive.

- Cost: Free premium account for 14 days that reverts to a basic WebEx Meetings free account that limits usage flexibility.

- URL: http://www.webex.com

- Pros: Excellent platform for live interactive sharing of your desktop and distance conferencing with other people.

- Cons: Lacks editing ability, audio output quality is low, and you must download the proprietary player in order to view a downloaded video file.

- Useful for: Virtual meetings and online collaborations. WebEx has the ability to record screen capture, but I would not recommend it as my first choice.

- Ease of use: Pretty easy

- Editing ability: Zero. Unless you were to download the proprietary .arf file and manipulate it in third-party software, you have no ability to edit what you have recorded within WebEx.

- Audio quality: Live audio is pretty good for meetings. Recorded audio is barely acceptable.

- Ease of sharing with others: Pretty easy. You get a choice of a streaming URL, or you can download a .arf file. Both are relatively easy to use.

APPLE FINAL CUT PRO (MAC ONLY)

- Cost: $299.99 with education pricing available to registered students and faculty

- URL: **https://www.apple.com/final-cut-pro/**

- Pros: Full flexibility to do everything you could possibly want to do with a video or film

- Cons: Very difficult for a new user to use immediately

- Useful for: All video and film editing. This is what the Hollywood professionals use.

- Ease of use: Not so easy. You definitely need formal instruction to use Final Cut Pro.

- Editing ability: Expert. All the editing ability you could ever want.

- Audio quality: Excellent (as long as you are using a high-quality microphone)

- Ease of sharing with others: Pretty easy

FREE VIDEO HOSTING WEBSITES

YouTube

URL: www.youtube.com

- Pros: Everyone uses YouTube. If you choose to make your video public, complete strangers can find and view your work.

- Cons: Your video will be surrounded by a plethora of other videos and ads, which can be distracting. Also, depending on the type of account you have, you may have a time or size limit on how much you can upload.

Vimeo

URL: www.vimeo.com

- Pros: A much better, ad-free user experience, in my opinion.

- Cons: Although you get 500 megabytes per week for free accounts, it can take a while for your video to get posted, depending on the type of account. There is a dramatic decrease in the probability of an unknown person stumbling upon your video. This may be a pro or a con, depending on your personal goals for the video.

Screencast

URL: http://screencast.com

- Pros: Screencast is owned and operated by TechSmith. This is the same company that provides Camtasia and Snagit. It provides excellent service and a smoothly operating site. The upload quality is great, you can choose your audience, and you retain all the rights to your content.

- Cons: With the free account, you get 2 gigabytes of storage and a 2 gigabyte monthly bandwidth allowance. If you need more, you will have to enroll in a paid account.

C Hardware

With audio and visual equipment, you can go as basic as a cell phone or as complex as a full production studio. One you probably have, the other probably not. Production quality is tied to the quality of your technology. However, many newer computer systems come complete with quality tools to record and edit audio and video. To create a presentation video, you will, at minimum, need a camera that can record your face or a computer that can record your screen. Both methods will require the ability to record audio.

Before we begin, it's important to note: I'm good with technology, but I'm not an expert, especially in the vast world of audiovisual and audio. However, I know what looks and sounds good, and I'm sure you do, too. Finding the balance between a quality product and the technology available to you is important. This section is merely here to give you a few pointers on where to go. If you want specific technology reviews, check out CNET at **http://www.cnet.com**.

If you don't currently own any hardware that is capable of making a video with audio, I promise you, there are options. Browse through your school's website, and find out if there is a computer lab available—I bet you there is, and I bet it's free to students. Many schools, especially those trying to build their online programs, now have low-cost laptop and tablet rentals for students. Depending on where you live, the public library is also a great resource for technology. I didn't have a home computer until I was a junior in High School (and yes people had home computers then). I went to the Fremont Main Public Library and reserved time on their computers to do my schoolwork. It was a dimly lit tiny room, which had four monster-sized off-white computers. Each computer station had its own little cubical, and you could reserve it for up to 3 hours at a time, completely free. Libraries have since updated their technology beyond what I used in 1995 and most likely have access to a web cam for users.

Today, I personally use a MacBook Pro to record and edit all my screen capture and digitally recorded videos. I'm not advocating Mac over PC here; I wouldn't want to start a war, but many times, students ask what I use to make my digital media. As with all things, you must find what works best for you and your budget.

Lastly, don't be shy about asking your friends. Why else have 1,200 online followers if you can't ask for a little help when you need it? They may say no, but they may say yes. You might be surprised how much people want to help out students. At some point, everyone has been a student and hopefully they remember how hard the struggle to succeed was.

 ## Audio

Don't forget that at least one human will need to listen to your video in order to grade it. Perhaps, depending on the class, lots of people will be watching your video. Keep this in mind when thinking about audio quality. Personally, I use a Blue Yeti USB Microphone—which is amazing! I found it at my local music store on sale for $99 just this year. It's perfect for recording voice overs, podcasts, vocals, and live music. Would I buy it to use once just for a class? No. As a student who is also a musician, singer, or budding podcaster, would I invest to use it for my other projects if I didn't already have a way to record quality audio? Yes! I've tried a couple of other USB microphones, and this has been my favorite so far.

Of course, you can use the onboard microphone embedded in your device or computer. It's better than nothing, but it usually will not produce a high-quality product. A USB microphone of any kind is fantastic, and there are many options out there that are very affordable. I will let you Google for those.

In addition to the audio equipment, you also need to have a good location to record quality audio. The room you're recording in should be quiet and free of possible distractions. If you live in a home with other people, let them know to keep it down and not come in until you're finished. Nothing's worse than being half way through your presentation and having a new voice come in and say, "Hey, pizza is here." You also want to be mindful of how much your space echoes. If you're in a space with little to nothing on the walls, you will have more echoing in your recording. This is a bad thing. Placing blankets on empty walls can alleviate this. Turn off your cell phone and anything else that might chime, ding or buzz. Put the cat outside and the dog in the garage. Even the hum of a fish tank filter can be picked up on good audio equipment and will be very distracting to the viewer. Although, if Fluffy the Fish is in a bowl, she can stay.

 ## Quick Tips for Shooting Video

Mind your background.

If you're recording video of anything other than your screen, you want to make sure you have an engaging background. It can be as simple as a blank wall in your bedroom or as complicated as your creative mind wants to make it. However, here are a few points to think about when choosing a background for your shots:

- Is there anything embarrassing behind you that you wouldn't want your instructor to see?

- Is the background you're using cluttered, messy, or packed with stuff? This can be very distracting to the viewer. Think about the impact your background will have on the viewer's ability to watch your video.

- Is there anything that you might sit or stand in front of that would look like it was shooting out of your head on camera? For example, you're standing in front of a super skinny lamp pole. It is directly behind you, but is much taller than you. To the camera, and thus your audience, the pole seems to be coming directly out of your head. Try to avoid this at all costs.

- -Does your background help to move the story along? This is a little bit more advanced, but if you are the creative type—you could have a lot of fun with this. Does your background help to tell the story of your video presentation? For example, I once did a video shoot for a story on how the human cardiovascular system works and used the Thunderhill Racetrack in Northern California for my background. I used one of the driver's engines as an analogy for the human cardiovascular system. Having an engaging background helped me to tell my story and make what I was saying a little more memorable.

Frame your shot.

The best option is to have someone else record you. That way, he/she can concentrate on it looking good while you concentrate on sounding good. However, this is not always possible. Usually aim for a straight-to-camera shot where the focus is on you and what you're saying; it is best to frame it so your face is clearly visible. I would pick one of these three types of shots for your presentation video:

1. Tight Shot: The camera is framed around you so that the bottom of the shot is just below your shoulders, or at breast height, and the top of the shot is a few inches above your head. With a shot this tight, you see more detail in your face and your expressions. Background is less important in this type of shot. However, you do want your background to be simple and not cluttered in any way.

2. Waist-Up Shot: This is your classic newscaster shot. The bottom of the screen will end at your waist and the top of the screen will end a few inches above your head. If you're recording with more than one person in the shot at a time, either sit on chairs or have the shorter person stand on a box to help even out any differences in height. This will help make the shot look cleaner. For background, you can use your presentation slides or any other adequate, non-distracting background that fits well with your video's theme.

3. Full Body: Firstly, if you choose to do this type of shot, go all the way. Make sure you can see your head and your feet in the shot. You don't want to cut yourself off at the knees—it just looks odd. Since you will be less of a focus in this type of shot, you want a really engaging background. Your background should be part of the story, since it will be filling up so much of the frame space. This is where you would definitely want to include images of your presentation slides or some other backdrop to help reinforce your message.

Secondly, since your entire body is being recorded, be mindful of how much you are moving. Move slowly and only when necessary. Recording quality movement shots can be tricky and relies heavily on the equipment you have, lighting, and how much help you have to record. However, I encourage you to experiment and see how you can create the best shots possible.

F Quick Tips for Computer Screen Recording

If you are not interested in including yourself in your presentation video, then doing a screen capture or screen cast is a great option. This is a type of project that will record your voice at the same time as your desktop screen. It is great for tutorials, as well. Take a look in the previous section, above, for screen capture software options.

For the best product, you want to use a software option that you can edit. Most students, especially when doing a long recording, will skip a line or forget something somewhere. Having the ability to edit can allow you to re-move those less-than-desirable portions and create a clean, well-produced product to submit for credit.

Audio is still very important, so check out the audio section, above. In theory, screen casting is just like recording with a camera: remove anything from your desktop or active windows that you may not want the world to see. It's probably time to delete all those cat videos anyway.

However, just in case you can't part with your desktop full of kitten pic-tures, WebEx, Quicktime, Snagit, and Camtasia will all allow you to record only one program or just a portion of your desktop. For example, if you are using PowerPoint to deliver your presentation slides, you can choose to share only the PowerPoint program with the screen capturing software. Then, your recording will just be your slides and your voice over, rendering your selfie collection safe.

Johanna Kubin Sardá

"Somewhere over the rainbow, way up high. There's a land that I heard of once in a lullaby." The words passed my lips as my vocal cords vibrated for my a cappella audition. The words rang true in my heart. On that stage, I could already imagine myself on an airplane, fighting for peanuts. I could feel myself walking in the streets of a big American city and into classrooms, chatting in English, learning new and interesting things. I was living a dream.

That song, as well as my academic record, curriculum vitae, and music experience allowed me the opportunity to leave my home country of Brazil to study for a semester abroad at Morehead State University. There, I studied with Dr. Roosevelt Escalante, a renowned jazz vocal coach, among other fantastic professors in the music department. Although my classmates and I were apprehensive about being able to communicate properly in English, I felt drawn to this opportunity and was going to explore it no matter what.

I arrived in Kentucky and was warmly greeted by Professor Frank Oddis, coordinator of the exchange program, and the University staff. From the airport to the dorms, I felt this was my path. Everything was ready for the new semester to begin. Then, homesickness and nostalgia hit hard. It is during those quiet moments when you have a clear and panoramic view of your life, memories flood your mind until you feel you can no longer move. You seem to hear the voice of your niece, the laugh of your father, and the words of encouragement from your siblings. You can even smell your mother's perfume in the air. You feel far away from everything you have ever known, but after the waves of longing subside, you are left with the strength to endure and push forward.

Taking risks to communicate with people outside of my comfort zone allowed me to feel at home in a foreign land. It allowed me to create lifelong bonds with people I would have never met otherwise, which is why

I consistently encourage others to experience life abroad. My English has improved a lot, and I will always carry those memories with me, wherever I go. In the end, dreams really do come true.

Johanna Kubin Sardá
Santa Catarina State University
Brazil

10 Short Films and Simple Animations

Short Films and Simple Animations

 a. Short Film Hardware

 b. Video Editing Software

 c. Animation Software

Today, many science classes are including short films and animations to allow students to express themselves. For example, I teamed up with the Center for Biological Diversity and launched a project where students created short films or animations of 30 to 90 seconds about the impact of human population on wildlife. Students were asked to showcase the Condoms for Conservation they were given in class. Yes, you read that correctly. I passed out 450 packages of condoms adorned with polar bears, plovers, and leatherback turtles all in the name of conservation. Don't believe me? Check out their program at EndangeredSpeciesCondoms.com.

Students took their tiny boxes of condoms, which had slogans like, "Don't Go Bare, Panthers Are Rare" and "Safe Intercourse Saves the Dwarf Sea Horse," and created amazing videos of their own design. The only requirement was the video needed to explicitly discuss how the global population negatively impacts species all across our planet. As the human population continues to skyrocket, so does the need for humans to have land, consumer materials, and food. Students could create this video alone or in teams. Once they had their final video or animation created, they needed to upload it to Twitter to share with the world.

It was such a unique opportunity for students to express themselves creatively. More instructors are using new media as a way to engage their classes in vitally important conversations. If you want to check out some of the videos, go to www.twitter.com and search for #CrowdedPlanet.

In this chapter, we will explore a new way for students in the sciences to communicate—through visual media. Technology has allowed users with every level of expertise to explore video and animation. I encourage you to poke around these new ways of expressing yourself. Personally, I have vastly more experience with short films than I do with animation. However, I have included easy quick-start ways for you to get your creative science project done and looking incredible.

Once upon a time, creating quality movies and videos was extremely expensive and required a lot of time, energy, and skill. Even though you still need all those things to make high-end productions for large scale screenings, it has become extraordinarily easy to create quick videos that can look polished with a minimal amount of effort. In this chapter, we will continue our video discussion from Chapter 9 and move deeper into the realm of video editing and animation. All pricing and websites are current as of this publication, but of course, please see the individual maker of each product for up-to-date information. All the reviews contained below are my personal opinions based on my experience with each product. If you have any additional feedback or find a wonderful resource I did not include, please consider letting me know at www.WritePresentCreate.com.

Short Film Hardware

Most people, especially for a short class project, will not be using actual film. However, most people still refer to shorts as short films or videos. In terms of hardware, you can use anything from the latest smartphone or a point-and-shoot digital camera to a high-end DSLR camera costing thousands of dollars. We are going to leave specialty and big budget cameras out of this discussion completely.

Creating videos has become extraordinarily easy. However, film and video quality is directly tied to the equipment you're using. You may already have a good option for small videos in your pocket right now or installed on your laptop. If not, there are a lot of options out there to help you create your video assignment. If you don't own any video recording equipment or software, then this is the perfect time to ask! Ask your instructor for advice, your fellow classmates for assistance, or even post a request on social media informing your friends that you're in need of video equipment or an extra pair of hands to help. You would be surprised what you can get when you have the courage to ask.

When choosing your equipment, think about your overall goal and where this video is going to be seen. Will it be seen only on a very small screen, like a phone or small tablet? If so, then go ahead and shoot with your phone. If not, be cautious. A few phones are capable of capturing enough detail to be displayed on a big screen without looking fuzzy or pixelated but not all. If you're unsure if your phone can do the job, experiment by taking a short video and displaying it on your target screen or ask a friend who is deeply knowledgeable about camera equipment and technology.

Additionally, will your video be viewed on a video hosting website, such as YouTube or Vimeo? If so, it's best to use a higher quality camera. This may or may not be your

> **DSLR**: Digital single lens reflex camera
>
> **Digital Media**: Any media that can be created, edited, stored, or distributed on electronic devices

phone, but you want to be able to display your final product at least at a minimum of 720 pixels.

For a class project, the easiest to use is digital media. Unless film is your passion, I would not recommend diving straight into trying to edit physical film. It can be done, of course, but it is more complicated than what will be discussed in this book.

Whether or Not to Tripod

Many newer cameras have the ability to smooth out the shake that occurs when any normal human holds a camera in his/her hands and attempts to record for any length of time. Zoom makes this shake even more noticeable. Again, it really depends on how sophisticated the technology you're using is. I always like to err on the side of using a tripod. A tripod also allows you to record yourself. If you don't have access to a legitimate tripod, you can also use anything that would be the appropriate height—dresser, shelf, or refrigerator.

 ## Short Film Software

Now that you have your fantastic video recorded, it's time to polish it up and share it with the world. In order to do that, you need to get it off your digital media storage card and into a software platform that can work with your file.

Just like with the hardware, film editing software ranges from free to expensive. It also ranges vastly from pretty simple to very complex.

By definition, editing is just taking what you recorded in your device (raw footage) and changing it in some way. This can be as simple as cutting off the beginning and the end or as complex as all the fancy effects in big-budget Hollywood movies. Here are my personal reviews of some of the most popular video editing software available:

QUICKTIME

- Cost: Free with updated Mac OS X Operating Systems

- URL: https://www.apple.com/quicktime/download/

- Pros: QuickTime will do very simple edits. QuickTime is good for trimming the front and/or end of your clip. Of course, you can also record in QuickTime. Depending on the goal of your final project, you could record and edit simple clips directly on your desktop/laptop using your web camera, onboard microphone, and QuickTime.

- Cons: You need a Mac in order to use this software, and there are not a lot of in-depth editing features available.

- Ease of use: Super easy

- Level of editing capabilities: Basic

Clip: A portion of a longer video or film

IMOVIE

- Cost: Free with updated Mac OS X Operating Systems

- URL: https://www.apple.com/mac/imovie/

- Pros: This is a great starter software. You can do lots of cool stuff with very little time invested in learning the program. You can trim, combine multiple clips, and add music, title slides, and lots of other cool stuff. As in other software programs, what you gain in ease of use, you loose in how many details you can manipulate. But for class projects, it's a perfect video editing tool.

- Cons: Your file size will be larger after editing, but if file size is not a concern, you don't need to worry about this. Additionally, you must have a Mac in order to use this software.

- Ease of use: Pretty easy

- Level of editing capabilities: Moderate

LIGHTWORKS

- Cost: Three options available – $24.99 per month, $174.99 per year, or $437.99 for a full license that never expires

- URL: **https://www.lwks.com**

- Pros: Works on all platforms, even Linux. This has everything you could possible want in a video editor. Professional films, like one of my favorites, *Pulp Fiction*, have been edited on Lightworks.

- Cons: With more control comes a steeper learning curve. You will need to spend some time learning how to edit your videos, but you will have all the editing control you would ever want.

- Ease of use: Not so easy

- Level of editing capabilities: Expert

WEVIDEO

- Cost: Edit and export up to 5 minutes per month for free, then $4 for up to an hour of exports per month

- URL: **https://www.wevideo.com**

- Pros: A lot of power with very little startup time required before you can produce edited videos. You can do Picture in picture, green

> **Picture in Picture**: Two distinct recordings are shown together simultaneously on one screen
> **Green Screen**: A technique that allows moving objects to be recorded in front of a solid-colored background, usually green, and an alternate background is added during the editing process
> **Ken Burns Effect**: Panning left to right and zooming in and out on still photographs

screen, and the Ken Burns effect. All browsers are supported, and you can even edit using your handheld device (tablet/phone).

- Cons: This is a really strong introductory video editing software that services all platforms. The only downside would be the monthly cost needed to export anything over 5 minutes per month.

- Ease of use: Super easy

- Level of editing capabilities: Moderate

MAGISTO

- Cost: Free for extremely short videos (generally 20 to 45 seconds). You must purchase a premium package, ranging from $4.99 to $9.99 per month, in order to get videos of 2.5 minutes.

- URL: http://www.magisto.com

- Pros: This is an automatic video editing software. Using what Magisto calls *Emotion Sense Technology*, an artificial intelligence software can detect the mood of your video by scanning the uploaded footage and creating post-production additions to elicit a targeted emotional response: happy, sad, exciting, etc. This software adds a lot of bells and whistles, based on a prescribed logarithm, to make your video really pop. Magisto is better suited for creating fast, engaging videos for friends and family. Since your ability to edit is almost completely taken away, I don't feel it is a great option for most academic projects. However, you never know! Magisto is available to you as a web-based application, as well as on iOS and Android devices.

- Cons: Based on the content, your project is funneled into a pathway that has been predetermined. You get a lot of flash with very little effort, but you loose almost all control of your final product. For example, Magisto decides which parts of your uploaded video are the most exciting. If you're doing a scientific presentation, this could be disastrous. Lastly, Magisto only allows you to create really short videos. Currently, two and a half minutes for premium clients whom pay by the month or the year.

- Ease of use: Super easy

- Level of editing capabilities: Basic

 ## Animation Software

Employing animations, whether as a stand alones or additions to originally recorded video, is an effective way to convey your message. I love animations, but I am not an animation artist. Having the skills to do great, original animations has a steep learning curve, but it is possible to make simple animations fairly easily with a little help. Below are a few websites I have used for some presentation-worthy animations. These websites will not land you an animator position at Disney Pixar, but they will help to make your presentation pop.

WIDEO

- Cost: Free

- URL: **www.wideo.co**

- *(side note: that is not a mistake, the URL extension for Wideo is in fact just a .co not a .com)*

- Platform: Web based—will work with PC or Mac

- Pros: Extremely easy to use, and you can include original content

- Cons: A decreased level of flexibility and control for creating original animated images, when compared to more complex programs, like Adobe After Effects

- Ease of use: Super easy

GOANIMATE

- Cost: 14-day free trial or minimum $39 per month

- URL: **http://goanimate.com**

- Platform: Web based—will work with PC or Mac

- Pros: Extremely easy to use, and you can add characters with voice

- Cons: If you use it past the 14-day free trial, you will need to purchase a monthly plan in order to have access to the videos you have created

- Ease of use: Super easy

VIDEO SCRIBE (WHITEBOARD VIDEOS)

- Cost: 7 days free trial at **http://www.videoscribe.co/freetrial**. Then, $29 per month, $198 per year, or $665 for a lifetime.

- URL: **http://www.videoscribe.co**

- Platform: Mac/PCs and iPad

- Pros: Create original hand-drawn whiteboard clips and animations. This is especially wonderful if you have access to touchpad technology. If you completely lack any artistic ability, video scribe does have some pre-loaded images for you to use, as well as a host of royalty free stock images, music, and fonts.

- Cons: It is not particularly cheap, so that can be prohibitive for students to use regularly. The 7-day free trial can get you through one project but really prevents most students from wanting to use it consistently.

- Ease of use: Pretty easy

ADOBE AFTER EFFECTS

- Cost: Relatively expensive to purchase, but you can do a 30-day free trial or purchase a monthly subscription, starting at $19.99. There is educational pricing available, and since this can be bundled with other Adobe products (Photoshop, Illustrator, etc.), the final price can range widely.

- URL: **www.adobe.com/products/aftereffects.html**

- Platform: Software available for installation on Mac OS X and Windows. Adobe After Effects is also available across all devices via the Adobe Creative Cloud.

- Pros: This is what the professionals use. Anything you want to do in animation, you can do with this program. Even if you don't have experience, but you have started to fall in love with animation, you might consider giving it a try. You may find a new passion!

- Cons: More flexibility and control means a more complicated program. It's not impossible for the new user to create a simple motion graphic in Adobe After Effects, but if you're brand new to animation, I would probably recommend a different option.

- Ease of use: Expert level

Tools for Academic Success

A How to Stay Engaged and Focused in Lecture

Sleep deprivation, stress, friends, family ... life! All these things can distract you from being present and engaged during lecture. However, the best time to start studying for an exam is during lecture. Nearly everything that will appear on the exam will be discussed in lecture. If you sleep or daydream through it, you are losing out on a valuable time saver and performance enhancer. Here are a few of my favorite ways to stay focused during lecture:

- Keep your body engaged through quiet tapping of a foot, knee, or finger. Nothing so prominent as to disturb your neighbor, but if you are feeling drowsy, this can help you stay alert.

- Take notes: See below for more information.

- Draw in your notes. If we are talking about the human heart, spend a little time drawing into your notes—even if the instructor has already left that slide. Try to draw it from memory. Even if you are not a talented artist, drawing and sketching is an excellent way to commit biological concepts to memory.

- Write exam questions during lecture. Especially write down those concepts the instructor says more than once. That is a dead giveaway that it will appear on the exam.

- Try not to have a really heavy, carbohydrate- or sugar-laden meal before coming to class. We all know the food coma all too well; circumvent this destructive measure by having a light meal before class instead.

TAKING NOTES

Don't fall into the trap of thinking you can just read the slides and be okay on the exam. I supply my lecture slides to students to remove the stress of getting all the information on them down on paper during lectures. However, this does not mean you get to skip the note-taking process. Taking notes in lecture keeps you engaged and actually counts toward your study time. Lecture should be an active exercise, not a passive one. The more engaged you are in taking notes, writing down questions to yourself about things you need to revisit, and making connections between concepts, the easier it will be to prepare for exams. High-performing students know that exam preparation begins on the first day of lecture.

 ## Studying and Preparing for Exams

BE THE PROFESSOR

Be the professor and create your own exam. The chances are high that you will have the exact same questions that they write, and you will already know the answers. Simply go through the material and create multiple choice questions based on the content. Not all content makes for good questions, so getting your mind into "exam mode" by writing questions and their answers will help you on test day. Once you create your exam questions, write the question on the front of a flash card and the answer on the back. You can now test yourself on whether you can correctly answer your exam questions. When writing these, always put them in a question form that will have only one answer.

BREAK UP STUDY TIME

On average, most people can be actively engaged in studying for 20 to 60 minutes before they start losing focus.

If you're not getting the grade you want and feel you deserve a better score on exams, look at your study habits. Many times, students try to sit down on their day off and study for four, six, eight, or maybe even ten hours straight. After all that hard work, they can't understand why they're not seeing the results. In these cases, it's not the level of work that is insufficient; it's the strategy.

INCORPORATE COLOR AND SMELL INTO YOUR NOTES

Use colored/scented markers. I recommend Mr. Sketch Scented Stix Watercolor Markers. I use the finer-tipped ones to create my notes. They are available in office supply stores and online—just Google the name. The more senses you involve with your studying, the better you will commit the information to memory.

MAKE YOUR SKELETOMUSCULAR SYSTEM HELP YOUR BRAIN

It may sound strange, but incorporating your skeletomuscular system into your study time can help you stay alert and commit information to your memory more efficiently. By doing some simple body movements while studying, you can increase focus and energy. Make sure to do something simple that has very little chance of injury, since your attention needs to be on your study material. Anything simple and repetitive works: squats, pacing back and forth, calf raises, slow stationary biking, leg lifts, using a balance ball for a chair, squeezing a stress ball with one of your hands, etc.

PARAPHRASE

In your own words, rephrase the material presented in the lecture slides. Paraphrasing, creating your own analogies, or coming up with your own examples ensures that you have a firm grasp on the material. Repeat back your paraphrased material in writing or verbally to yourself or a friend.

GROUP STUDY

Pair up with someone who cannot only quiz you, but also encourage you to go into detailed explanations. Don't just read your notes to each other; present the information in your own words, just like you would teach a student in a class or in a tutor session.

In your study group (two to four people usually works best), have each person develop five questions for each chapter/unit. Combine them and create a practice exam!

Quiz each other prior to the exam—that morning, the night before, etc. Group/pair studying really works. Teaching a topic to another person is one of the greatest methods for getting information into long-term memory. This is a higher stage in learning theory above that of rote memorization. (You want to avoid rote memorization as much as possible, since this does not allow you to really know the material; it usually only commits the information to short-term memory).

Theory: A coherent set of repeatedly tested and widely accepted statements or principles devised to explain or predict natural phenomena

Acronym: A new word formed from a set of words by using the first letter or first groups of letters from each word

MEMORY GAMES: MNEMONICS

Using acronyms, acrostics, rhymes, jingles, stories, and songs can work really well for some students. For more information on how to create your own mnemonics, check out: http://college.cengage.com/collegesurvival/wong/essential_study/6e/assets/students/protected/wong_ch06_in-depthmnemonics.html

VISUALIZATION

Practice visualizing the diagrams, charts, concept maps, and summaries you create. Without looking at your study materials, try to recreate them visually in your mind. If during your attempt at visualizing your materials you can't quite fill in all the details, go back and look at them again. Repeat this process until you can clearly visualize, with details, your study materials in your mind's eye. Then, during the exam, you can mentally flip through your notes.

CONCEPT MAPPING

Concept mapping is a graphical representation of data or related information. There are many styles to choose from, including matrix, flow chart, etc. Check out this short video on how to create your own concept maps:

http://www.youtube.com/watch?v=vuBLI6ijHHg

C Additional Tools for Success

TIME MANAGEMENT

Use a calendar to schedule your time and remind yourself of important deadlines.

- At the beginning of each semester, take some time, gather all your syllabi/schedules, and enter in every deadline, exam, and important date.

- Almost all electronic calendars (Google, Outlook, etc.) will allow you to set reminders for events. Set a reminder for exams, papers, etc., far enough in advance so if you happen to forget, your calendar will remind you in enough time to do something about it.

- Managing your time appropriately decreases your stress and makes you more efficient.

DEALING WITH TEST DAY ANXIETY

Everyone deals with anxiety to some degree on test day. Here are some tips to help decrease that anxiety as much as possible:

- Don't drink coffee/caffeine: Caffeine stimulates the central nervous system and can actually create more anxiety, nervousness, shakiness, dehydration, and a lack of focus. Although studies have sometimes correlated caffeine with an increase in performance, there are also many more studies that support the negative effects of caffeine on people who are already in a state of anxiety and nervousness—such as during an exam. In addition, since caffeine is a diuretic, it can make you dehydrated, which can in turn cause a decline in mental clarity and alertness.

- Get to the exam room early, at least 15 minutes before the exam begins. Nothing sets a student up for failure like the stress added to running late for an exam. Plan ahead, and be there ridiculously early. That way, you can find your favorite seat, relax, and gently go over your notes. In this way, you are mentally prepared once the exam begins.

- Eat breakfast. Many students swear they're not hungry before an early exam. At minimum, try and eat a little protein and a bit of healthy carbohydrates. Having a little something in your stomach will give your body the energy it needs to perform at its optimum ability during the exam.

- Get a full night of sleep before the exam. Cramming and all-nighters don't work. Studies have shown that an all-nighter will generally result in lower grades, a decreased ability to focus, depression, and disruption of normal sleep patterns for up to two weeks. Don't believe me? Have a look at the study published in *Behavioral Sleep Medicine* (Thacher 2008): "University Students and the 'All Nighter': Correlates and Patterns of Students' Engagement in a Single Night of Total Sleep Deprivation."

- Many students find it helpful to have an aisle seat. This can make you feel like you have more space and easy access to walkways.

- Find a favorite pencil—one you really love—and use it during your study time and test time. This may trigger some of the material you reviewed during your study sessions and make you feel more at ease in the exam room.

THE QUALITY OF YOUR HEALTH IS DIRECTLY CORRELATED TO THE QUALITY OF YOUR ACADEMIC PERFORMANCE

Food is fuel. The quality of fuel you put into your body, and thus your brain, will directly impact the quality of your performance. Think of your body and your brain as a high-performance car. You wouldn't put low-quality gas into a Lotus or Ferrari, would you? Quality fuel for you means lean proteins, with a full complement of amino acids, complex carbohydrates, and healthy unsaturated fats. Try to minimize excessive sugar, saturated fats, and caffeine—all of which have been proved to derail peak performance.

Courtney Ables

If you're the best person in the room, you're in the wrong room. I learned this valuable lesson my freshman year of college after inadvertently enrolling in an upper division sociology course. That first day was overwhelming as a freshman, and once I found out my error, my first instinct was to run for the door and never look back. Before dropping the course, I went to discuss my options with my school counselor. She advised me to stay and give the course a try. Although I was nervous and  apprehensive, I took her advice and was immensely grateful I did. Although being the only freshman in the course felt lonely at first, I took great pride in knowing I was learning at a more rigorous level than my peers. Soon, I was able to connect with my fellow classmates, and they went out of their way to show me the ins and outs of campus life. They even told me about a secret parking lot! By taking a risk, I was able to have a rewarding experience and discover the path that would successfully take me into my future career as a sociologist.

Courtney Ables
Ohio University

12 Copyright Guidelines for Use in Higher Education

Copyright Use in Higher Education

    a. What is Copyright?

    b. Selected Types of Licensing

    c. Sources of Free and Low Cost Legal Images

 ## What is Copyright?

Although the internet provides a seamlessly unending plethora of writings, drawings, videos, and graphics, they were made by humans. Therefore, in the United States, they will have a varying degree of legal protection. Copyright law protection applies to a vast array of works: artistic, scientific, and literary. This legal protection impacts who can use and distribute a creator's work. As web-based digital media becomes

> **Torrenting**: A practice of downloading files from more than one server or user that may or may not have permission to share the content

more and more pervasive, copyright protection becomes an even more important topic to discuss. Depending on the situation, you may be prosecuted for breaking U.S. copyright law if you use or distribute protected work that has not received the appropriate permissions. Of course, this is usually only applied to those who intend to income from the illegally used work but not always. In the wake of Napster and shared music, some users were, in fact, taken to court over illegally downloading. Today, Torrenting is still popular, even though it does involve illegal downloading of copyrighted material.

During your education projects, the chances of you being prosecuted for using an image you copied and pasted from a Google search without permission is pretty miniscule. However, there is a right way and a wrong way to use media in your academic work. It doesn't matter if it's audio, visual, or written—every piece of work has its own level of copyright protection. Just like it's good to get into the habit of only using reputable resources, it's good to begin using material legally in your projects now and in the future. For all the information you may or may not want to know, start at the source by visiting www.copyright.gov.

Copyright protects works that are "fixed in a tangible medium of expression," which means that ideas are not protected under copyright law. Here are just a few specific examples of creative works that would be eligible for copyright protection:

- Movies and Videos: Scripts and Productions

- Music: Recorded Music, Sheet Music, Orchestras, Operas

- Writings: Books, Blogs, Magazine Articles, Newspaper Articles

- Images: Paintings, Photographs, Designs

- Dance And Theatre: Choreography, Plays

- Technology: Devices, Software Code, Hardware Components

Copyright covers both published and unpublished works and can last decades, even millennia, after the author dies. This is provided for by the United States Constitution in order to protect creators of original works. Copyright protects what is referred to as "tangible works." Tangible works are things you can hear, see, read, touch, taste, or smell. Copyright protects books, architecture, songs, plays, software, poetry, and art. It does not protect facts, ideas, protocols, or methodologies. It also does not inherently cover inventions, discoveries, phrases, logos, or symbols. That is where patent and trademark law come in, but that is for another book.

In our current technologically driven society, where we can screen capture, lift, copy, and paste images, music, and videos: it is up to us to be respectful of the artist. If the artist says, "please, use my art and pay me what you deem fitting," then you do. If the artist says, "I spent hours, days, weeks, maybe even years on this piece. Please be respectful by not taking without giving," then I urge you to be respectful. Art, in all its forms, is a personal expression, and we all need to be respectful others wishes around their work.

> **Patent**: Legal exclusive rights granted to an inventor for an invention

COPYRIGHT AND FAIR USE

According to Stanford University, fair use is the reproduction of copyright material for a limited and transformative purpose. Such uses could be to teach a topic, participate in scholarly activities, criticize a work, create a news report, comment on a work, or create a parody. For example, Weird Al Yankovic (http://weirdal.com) has made an entire career by creating pop-culture parodies. We see them all the time, and they are made possible through the fair use exemption.

The definition of what is and what is not fair use is left ambiguous by design. What is considered "transformative" is difficult to distinctly define and, thus, creates an abundance of possible interpretations. For further information on fair use and its role in copyright law, I would highly suggest you check out fairuse.stanford.edu.

Fair use is extremely helpful in the classroom. However, as with all laws, it is not a straightforward "everything is fair game" exemption. The following two laws discuss fair use in the academic setting.

THE CLASSROOM USE EXEMPTION

This exemption to the copyright law allows instructors and students to display, discuss, and use copyrighted works in a not-for-profit academic setting. But, there are restrictions.

To use this exemption, you must be physically in a classroom devoted to instruction at a non-profit educational institution. If you're taking an online course or attending a for-profit school, this exemption cannot be used, and you will need to have explicit permission from the copyright owner to display, showcase, or perform their work. Additionally, the Classroom Use Exemption does not apply to reproductions. Reproductions involve photo-copying handouts of a copyrighted work, burning DVDs for the instructor or classmates, or distributing any unlawfully made copies in any way.

THE TEACH ACT

When offering online examples of copyrighted material, things get even trickier. In 2002, President Bush signed into law the Technology, Education and Copyright Harmonization, or TEACH, Act. This provided for some additional exemptions for online learning at accredited institutions of higher education. For more in-depth information on the TEACH Act, please visit *copyright.com/Services/copyrightoncampus/basics/teach.html*.

 ## Selected Types of Licensing

Legal rights around artistic works are deeply intricate, and this guide is not meant to educate you on all copyright and patent law. It is simply meant to give you just a little taste of what is out there when you are creating a project. Maybe you're already pre-law or a political science major, but if you find this chapter interesting, I highly encourage you to keep diving deeper into issues of law. You may even find a new passion that will lead to an exciting path of learning and exploration.

ROYALTY FREE (RF LICENSE)

Royalty free is not truly free. A royalty free, or RF, license, was obtained by someone, at some point, by paying a fee to the creator. After that agreed-upon fee was received, the creator granted the payer the right to allow others, sometimes for an additional fee, to use the artistic images, music, videos, or animations without paying the creator any additional money, otherwise known as a royalty. In other words, royalty free images are free to use without paying additional royalties, but they still must be purchased by the user or obtained from a source that provides royalty free images to their customers. Royalty free images are usually sold for a flat fee, which allows the purchaser to use the image in almost all purposes.

RIGHTS MANAGED (RM LICENSE)

A rights managed, or RM, license, is much more restrictive than an RF or creative commons (CC) license. Instead of paying a flat fee for the work and getting to use it for almost any project you could conceive of, a price is created for you based on your plan. The seller will place restrictions on it, such as restrictions regarding which countries it can be distributed in, length of time, and mediums, among other specific limitations.

CREATIVE COMMONS LICENSES

Creative Commons is a non-profit organization with a mission to help make creative works already in existence become more accessible to artists of all types. A portion of the original vision of the U.S. copyright law was to help foster creativity and ingenuity, not stifle it. Just as in science, creative arts work best when we can build upon the good work that has come before us. Of course, there should always be protection for the creator, but Creative Commons gives creators the choice of how they wish to share their work with the world.

Creative commons (CC) licenses work alongside U.S. copyright laws to help complement them, not replace them. Creators still own their works with a creative commons license. The non-profit group Creative Commons offers six different licenses, based on the needs of the creator.

As taken directly from Creative Commons, its mission is stated as, "Creative Commons develops, supports, and stewards legal and technical infrastructure that maximizes digital creativity, sharing, and innovation." For more information on Creative Commons and its resources, please check out creativecommons.org.

STOCK PHOTOGRAPHY, VIDEO, AND MUSIC

Stock images, video, and music are generally provided by a company that has pre-cleared all copyright permissions before posting it to be purchased by you. Most stock work is reasonably priced, depending on the company, and will most likely be sold for a flat fee with no royalties. Purchasing stock work does not mean you own it, it only means you have permission to use it.

PUBLIC DOMAIN

Public domain creative works are tangible expressions that are not protected by any intellectual property laws, such as copyrights, patents, or trademarks. Since public domain works are, by definition, owned by the public—there is no need to obtain permission from the original creators to use it. Although no one can ever own public domain work, there is a small chance that it would be protected under copyright law if an artist made an original new piece of work using multiple public domain works. Then, the new piece could be copyright protected, although the originals would stay public domain.

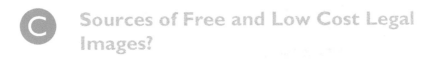

C Sources of Free and Low Cost Legal Images?

Below are some great sources to get free and legal creative works that are perfect for your project. This is not an exhaustive list, but it will definitely get you started.

creative commons search

URL: http://search.creativecommons.org
Notes: A portal to a plethora of images, video, and music that is available for your use under the creative commons license.
Types of work available: Images, video, and music

deposit photos

URL: http://depositphotos.com
Notes: Very high quality images with a wide range of topics
Types of work available: Photos, vector images, and videos

ccMixter

URL: http://ccmixter.org
Notes: Excellent for samples of vocals and electronic music
Types of work available: Music

pixabay

URL: http://pixabay.com
Notes: Array of subjects available, from animals to travel, in high clarity resolutions
Types of work available: Photos, vector art, and illustrations

URL: **http://www.loc.gov/pictures/**
Notes: Perfect for historical pictures used with an American lens.
Types of work available: Images

URL: **http://www.nih.gov/about/nihphotos.htm**
Notes: This is an excellent resource for images related to biology, health, and medicine
Types of work available: Images and videos

STUDENT BIO

Enrico Miguel Thomas

"Do you know why you're here? You're being charged with a felony!" Those words rang through my exhausted mind like church bells through a silent square. As I looked up at this man, strong and powerful, towering over me, my life began to flash across my mind's eye. A life of struggle that had left me filled with anger and bitterness. My memories flashed back to my childhood, where, at 3 years old, I was scalded by my father in a shower so hot, it left permanent third-degree burns all over my body. Burns that made me feel outcast and lost. Burns that were so deep, they colored my entire perception of the world. But, I always had my art. Drawing had gotten me through the toughest of times. Putting pen to paper had given me solace throughout my life, even as I was living in a homeless youth shelter, which was part of the reason I was there in that courtroom, charged with a violent crime I didn't commit. But who would believe me? I felt like I was nobody, until Kevin Sylvan, one of New

York City's top criminal lawyers was standing in front of me. Kevin was volunteering as a court-appointed attorney and was assigned to me that day, one of the luckiest days of my life. Kevin believed me, believed in me. Like a skilled surgeon, he was able to excise me from this disaster, and I was set free. I returned to the youth shelter determined not to let Kevin down. Communicating through images was where I felt most at home, so with the help of many others, I was able to receive counseling for my anger and threw myself into drawing. Communicating with the world through art has brought me to today. Today, I hold a Bachelor of Fine Arts Degree from Pratt Institute, one of the worlds top art colleges. Today, I work with Sharpie U.S. sharing my art with the world through banner ads and commercials, which have been positively reviewed by the New York Times. Today, I know it was communicating with others—through my words and through my art—that allowed me to not only live, but thrive.

Enrico Miguel Thomas
EnricoMiguelThomas.net
Pratt Institute, New York City

Image Credits